The Food-Energy-Water Nexus 2017

Topical Conference at the 2017 AIChE Annual Meeting

Minneapolis, Minnesota, USA
29 October – 3 November 2017

ISBN: 978-1-5108-5788-9

Printed from e-media with permission by:

Curran Associates, Inc.
57 Morehouse Lane
Red Hook, NY 12571

Some format issues inherent in the e-media version may also appear in this print version.

Copyright© (2017) by AIChE
All rights reserved.

Printed by Curran Associates, Inc. (2018)

For permission requests, please contact AIChE
at the address below.

AIChE
120 Wall Street, FL 23
New York, NY 10005-4020

Phone: (800) 242-4363
Fax: (203) 775-5177

www.aiche.org

Additional copies of this publication are available from:

Curran Associates, Inc.
57 Morehouse Lane
Red Hook, NY 12571 USA
Phone: 845-758-0400
Fax: 845-758-2633
Email: curran@proceedings.com
Web: www.proceedings.com

TABLE OF CONTENTS

Some abstracts in this book refer to images that were not made available by the publisher

505003 Introductory Remarks by Nada Anid

Abstract:

PAIC aims to identify, respond to, and communicate public-policy issues at local, national, and international levels. It also aids AIChE members in learning how to communicate better with government and the public. In light of this mission, the goal of the Town Hall meeting is 1) to share the findings of three PAIC Task Forces and get the membersâ€™ input on each and on other matters of importance to the AIChE membership and 2) to consider the best means of communication to enhance the standing of the AIChE community and express its stand on policy matters.

Copyright © American Institute of Chemical Engineers. All rights reserved.

504000 Food-Energy-Water Issues

Abstract:

The overlapping needs and technologies for food, energy, and water call for contributions from many people around the world, but chemical engineers bring especially valuable expertise. This briefing will summarize the issues and preview the topical conference that is taking place this week.

Copyright © American Institute of Chemical Engineers. All rights reserved.

504002 Advanced Manufacturing

Abstract:

New â€œAdvanced Manufacturingâ€ approaches are expected to impact the production of chemical products significantly. Recent advances include process intensification, manufacturing for sustainability, new sensors and actuators, data science, and high-level simulations. We will discuss what advances and further developments need to be communicated, why they should be adopted, and how they can be communicated effectively.

Copyright © American Institute of Chemical Engineers. All rights reserved.

504001 Climate-Change Review and Adaptation

Abstract:

The PAIC is leading an AIChE review of climate change and our policy on it. AIChE members are participating in a series of online Engage discussions, each introduced by an article posted on ChEnected. By drawing from the knowledge, dedication, and passion of AIChE members, AIChE can shape a Climate-Change Policy that captures membersâ€™ concerns while helping our profession to grapple with this issue.

Copyright © American Institute of Chemical Engineers. All rights reserved.

505002 PAIC Town Hall

Abstract:

In this Town-Hall-style, moderated discussion, the emphasis will be on identifying other critical public-policy issues where AIChE and chemical engineers can have impact. Attendees are encouraged to submit questions beforehand to the organizers by email or at the session. The goal is to shape action paths and find participants who are interested in engaging with each other and the public on these issues.

Copyright © American Institute of Chemical Engineers. All rights reserved.

506141 Into Hot Water: Utilizing Thermal Distributed Energy Resources to Improve Grid Reliability

Abstract:

Combined heat and power (CHP) technology along with grid-integrated water heaters (GIWH) are underutilized thermal distributed energy resources. They can be implemented on a large scale to provide ancillary services for the grid, including load shifting, frequency regulation, and improved grid reliability. The market is ripe for CHP and GIWH to be rapidly implemented on a large scale. This paper discusses the legislative and technical background of CHP and GIWH, the way these technologies provide benefits to a rapidly modernizing smart grid, and concludes with potential policies on the state and federal level that would support greater use of thermal distributed resources. Three policies are discussed in detail. The first of these is to expand the Department of Energyâ€™s CHP Technical Assistance Partnership program to facilitate knowledge and implementation of a shared ownership model for CHP systems between utilities and an institution that will utilize the energy produced. Next, states should encourage utilities to create incentives for consumers to install GIWH to be used in utility programs; these incentives should focus in particular on assisting low-income households to purchase GIWHs. Last, states should set goals to have all utilities that function on retail markets to adopt time-based pricing structures for electricity.

Copyright © American Institute of Chemical Engineers. All rights reserved.

506142 Turning the Tide: Policies to Advance Saltwater Desalination in the United States

Abstract:

Although water remains one of the worldâ€™s most abundant resources, most of the water resides in the ocean. With an already limited supply of freshwater, the increasing global population, changing climate, and worsening droughts are placing additional strains on the worldâ€™s water supply. Currently, several regions of the U.S. are experiencing some level of water shortage, and while the majority of the affected states reside in the West, the effects are felt nationwide. Although both water conservation and reuse remain key components of the national water portfolio, these practices alone will not sufficiently accommodate future water demands. Desalination, which removes salts and other suspended solids from seawater and brackish water, is a technique widely used around the world to convert high-salinity water into water suitable for human use. The unique benefits of seawater desalination remain to be recognized in the U.S., and since the options for augmenting the current water supply are limited, seawater desalination should hold a distinct place in the U.S. water portfolio.

Copyright © American Institute of Chemical Engineers. All rights reserved.

506143 Microgrids for the Macrogrid: Advancing Community Microgrids for Grid Modernization

Abstract:

Global trends - the digital age, prospect of cyberwarfare, climate change, consumer preferences â€" are irreversibly changing the supply and demand for electricity as well as the risks that threaten access to electricity. The existing grid was designed to serve a different kind of society and is ill-suited for meeting new challenges. A modern electricity system will not only have more actors, but will need to be fundamentally different in how it dispatches resources, communicates information, responds to grid disturbances, and plans for the future. A modern grid is decentralized, flexible, adaptable, and smart (i.e. automated, digitized, self-healing). By these characteristics, electricity access becomes reliable, efficient, and resilient. Microgrids add both resilience and flexibility to the system and policy initiatives can encourage them through technology-agnostic market rules and economic forces.

Copyright © American Institute of Chemical Engineers. All rights reserved.

510175 Small Scale Shock-Proof Biogas Digesters

Abstract:

Affordable, easily maintained biogas digesters have the potential to greatly increase the standard of living in several key communities around the world. From food scraps and animal wastes, these reactors can output high-quality fertilizer in addition to natural gas that can be used for heating, cooking, etc. Designing an assembly guide with the key features of a safe biogas reactor is key to developing this technology in the communities in question; particularly communities with limited access to education or manufacturing capabilities. Development of this assembly guide requires implementation and development of a small â€œtestâ€ reactor, including various mechanical systems, embedded electronics, and other features that would allow a more comprehensive and thorough understanding on the feasibility and overall maximum output possible by this kind of reactor. Secondary research would strengthen this knowledge base and further increase the ability for these reactors to be beneficial to their intended users, this research includes lab-scale experimentation on the growth rates of critical bacteria, determining decomposition rates of certain safety features, and detailing how other physical stimuli may result in differing output rates of the reactor. The presence of a simple, customizable biogas reactor in an impoverished community could potentially raise the standard of living for the surrounding populace by eliminating excess food waste, producing high-quality fertilizer for crops, and supplying a steady stream of clean natural gas for heating or cooking.

Copyright © American Institute of Chemical Engineers. All rights reserved.

506138 Computing at the Nexus of Food, Energy, and Water

Abstract:

In coming decades, the world population is projected to grow significantly increasing the demand for food, water, energy, and other resources. Furthermore, these resource challenges may be amplified due to climate change and urbanization. In addition, Food, energy and water (FEW) systems were traditionally analyzed and planned independently to address the challenges of population growth, climate change and urbanization. However, such piece-meal approaches (e.g., bio-fuel subsidy, fertilizers in agriculture) to solving problems in one system (e.g., energy, food) led to unanticipated harms to other systems (e.g., food price increase, water resource depletion and degradation). Thus, understanding the interdependent and interconnected nature of food, energy, and water systems (FEW nexus) is a societal priority.

Computing is crucial for understanding the problem, the interconnections, and the impacts withing FEW nexus. It is also needed for monitoring a variety of Earth resources (e.g., agriculture fields, fresh water lakes, energy needs for cooling or heating, etc.), and trends (e.g., deforestation, pollution, etc.) for timely detection and management of risks, such as impending crop failures and crop-stress anywhere in the world. It is also needed to reduce waste and to improve efficiency, e.g., amount of water and energy needed to produce food.

Computing success stories go beyond the cyber-infrastructure for simulations (e.g., GCMs, AgMIP) to include precision agriculture and GEOGLAM. Precision agriculture uses cyber-physical systems and data science to increase yield, and reduce fertilizer and pesticide runoffs. The Global Agricultural Monitoring (GEOGLAM) , an international system, uses remotely sensed satellite imagery to monitor major crops for yield forecasts to enable timely interventions to reduce disruptions in global food supply.

However, the FEW nexus presents new challenges and opportunities for computing. For example, data science methods need to not only re-examine assumptions such as non-stationarity (e.g., climate change) but also address nexus challenges such as high cost of false positives, (social) feedback loops, and multiple spatio-temporal scale.

Copyright © American Institute of Chemical Engineers. All rights reserved.

505270 Circling the Nexus

Abstract:

Growing support for circular economy approaches in the public and private sectors is drawing attention to the food-energy-water nexus. In the United States, the Materials Marketplace is an award-winning regional and national platform to facilitate company-to-company industrial reuse of all kinds. Through the cloud-based platform, traditional and non-traditional industrial water, energy, food and waste streams are uploaded and proactively matched with new product and revenue opportunities, ultimately enabling the culture shift to a circular, closed-loop economy. In addition to diverting waste from landfills, these recovery activities generate significant cost savings, energy savings, and create new jobs and business opportunities.

Copyright © American Institute of Chemical Engineers. All rights reserved.

492541 Limits to Growth and Global Sustainability of Food-Energy-Water Nexus

Abstract:

A mathematical representation of the major dynamics of the World, inspired by the "Limits to Growth" model[1] constructed in the 1970s, has been under construction for approximately the past ten years[2]. The mathematical representation includes an ecosystem with food-web model, a human population, industry, energy generation, a macro-economy, and rudiments of a legal system with private property, taxation, and regulation. The work was started in 2004 at the U.S. Environmental Protection Agency under the supervision of Dr. Heriberto Cabezas. Details of the model can be found in the scientific literature[3]. Around 2010, the project was transferred to the Vishwamitra Research Institute led by Dr. Urmila Diwekar in Chicago, Illinois, USA where the work has proceeded under the joint supervision of Dr. Urmila Diwekar and Professor Heriberto Cabezas from the Peter Pazmany Catholic University. Several publications have resulted from this research effort[4],[5], two of which are cited here. The mathematical model now includes water also making it a system where food-energy-water nexus can be studied and is developed sufficiently so that scenario analysis shows that consumption increase predicted for the future makes the compartments, i.e. species, of this model unsustainable. However, complex network analysis offers a pathway to analyze various versions of this model. The value of the model is it provides the conceptual machinery "steering" the system away from

---

[1] Meadows, D. H., Meadows, D. L., Randers, J., & Behrens, W. W. (1972). *The limits to growth*. New York, 102.

[2] Cabezas, H., Pawlowski, C.W., Whitmore, H.Wm., & A.L. Mayer, "On the Sustainability of Integrated Model Systems with Industrial, Ecological, & Macroeconomic Components," Res., Cons. & Recycl., 50, 122-129 (2007).

[3] Kotecha, P., Diwekar, U.M. and H. Cabezas, "Model Based Approach to Study the Impact of Biofuels on the Sustainability of an Integrated System," Clean Tech. Env. Pol.," 15 (1), 21-33 (2013).

[4] Benavides, P.T., Diwekar, U. and H. Cabezas, "Controllability of Complex Networks for Sustainable System Dynamics," J. Complex Networks (2015), published on-line. DOI: 10.1093/comnet/cnu051.

[5] Doshi, R., Diwekar, U., Benavides, P.T., Yenkie, K.M. & H. Cabezas, "Maximizing Sustainability of Ecosystem Model through Socio-Economic Policies Derived from Multivariable Optimal Control Theory," Clean Tech. Env. Policy, 17 (6), 1573-1583 (2015). DOI: 10.1007/s10098-014-0889-2.

Copyright © American Institute of Chemical Engineers. All rights reserved.

492950 Meeting Challenges in a Land Constrained Solar Economy: Co-Production of Energy and Food Using Photovoltaic Systems over Farmland

Abstract:

The main purpose of this presentation is to demonstrate that contrary to the popular belief that ample land area is available to install photovoltaic modules, in an economy mainly supported by solar energy, there will be severe land availability constraints in many regions of the world—including many states in the USA. This will inevitably lead to competition for land between food and energy [1]. *We therefore propose several innovative PV systems that will allow for the co-production of food and energy from the same unit area of farmland thereby allowing local energy needs to be met by local supply of photons.* This allows us to overcome feedstock (land) and transport (long-distance transmission) limitations for electrical generation that would otherwise be difficult to overcome in a solar economy.

A recent study by MacKay showed that the energy consumption per unit area in densely populated countries such as the UK, Germany, Republic of Korea, and Japan is approaching the energy generation per unit area of photovoltaic farms [2]. This suggests that a transition to a solar powered economy in these nations would require deployment of solar on a large fraction of the nation's total land area, which will directly compete with land requirements for other needs (e.g. agriculture). Many other countries, including the United States are also trending toward this scenario [2]. A recent study by National Renewable Energy Laboratories confirms the large land area requirements of solar farms [3].

Using the available data on current U.S. energy consumption and land availability [4]-[5], we have estimated that as many as 31 states may be unable to meet their needs without importing energy long distances from other states or offshore. This scenario arises because much of the open land in the U.S. is devoted to cropland (21.5%), pasture land (32.3%), and forest land (30.4%). Only 3.6% of the land falls in the category of miscellaneous/other use land [5].

As it should remain a priority to preserve our native forest and wildlife areas, one of the greatest opportunities for increasing the availability of land for solar energy generation is over farmland. These land areas are often easily accessed and surround population centers. However, if PV is to be successful deployed over farmland, novel systems will need to be designed that do not negatively impact crop production. As most photosynthetic activity requires only the visible portion of the spectrum (400-750 nm), these wavelengths can be allowed to past through to plants while the IR portion of the spectrum is harvested as electricity. It is also clear that plants can go relatively extended periods of time without solar irradiance (night) and survive extended periods of reduced lighting (cloudy days) without significant impact to crop production so it may be possible to harvest a portion of the visible portion as well.

Figure 1 shows several possible configurations for harvesting solar energy over farmland. Fig. 1 (a)-(b) are systems that could operate optimally using current industrially-proven solar absorbers (e.g. silicon). Fig. 1 (c)-(f) shows systems for which light splitting would result in even higher performance by optimizing the band gap of the absorber for the portion of the spectrum it is absorbing. Fig. 1(a) is the simplest case in which south-facing PV modules are

altered such that every other cell is left open or replaced with a light diffuser. In this system crops never experience shading for very long as the shadow will be constantly moving underneath the panel. Fig. 1(b) is the case in which single- or dual-axis tracking panels are used. Again, the shadow will move throughout the day. In the event that crops cannot tolerate extended periods of shade, these panels can also be patterned and be made twice the size to decrease the time intervals of shading, though the total time each plant would spend in the shade over the course of the day would be the same.

Fig. 1(c)-(f) are systems based on industrially available dichroic mirrors that split the spectrum [6]-[7]. These offer the advantage of concentrating the light for higher efficiencies. As a downside, diffuse portions of the spectrum may not be harvested efficiently. In Fig. 1(c), a tracking parabolic trough system is used in which the IR portion of the spectrum is reflected to a PV cell of appropriate band gap while the visible light passes through. As an option, a thin bifacial PV cells may be placed at the back (or a pattern thereof) to harvest a portion of the visible light and light scattered back from the crops. In Fig. 1(d), the system is non-tracking for mechanical simplicity and directs much of the IR spectrum to bifacial PV cells of appropriate band gap. During non-growing seasons, PV panels could be inserted in place of the transparent support. Fig. 1(e) depicts a tracking, long-pass system in which the visible light is concentrated and collected in fiber optic cables and then redistributed to plants underneath (and potentially other uses such as lighting. Fig. 1(f) depicts a non-tracking, long-pass system in which visible light is again collected in fiber optic cables and then redistributed to plants underneath, while the IR passes through to be collected on PV of appropriate band gap underneath. Promising preliminary calculations have been completed and are on-going to determine how much power can be produced and the optimal band gaps for each system.

The choice of which system to use will depend on how the crops respond to different shading characteristics and how much light deprivation and which portions of the spectrum they can tolerate being deprived of. It will also depend on the relative percentage of diffuse to direct light in a geographic location as the systems in Fig. 1(a)-(b) are better suited for areas with more diffuse light. We explore these options and make some general conclusions on system selection.

References

[1] Solar Projects Sow Tension, Wall Street Journal, Wall Str. J. (2017) A3.

[2] D.J.C. MacKay, Solar energy in the context of energy use, energy transportation and energy storage., Philos. Trans. A. Math. Phys. Eng. Sci. 371 (2013) 20110431. doi:10.1098/rsta.2011.0431.

[3] S. Ong, C. Campbell, P. Denholm, R. Margolis, G. Heath, Land-Use Requirements for Solar Power Plants in the United States - Technical Report NREL/TP-6A20-56290, 2013.

[4] State Profiles and Energy Estimates, (n.d.). http://www.eia.gov/state/.

[5] Major Land Uses, (n.d.). https://www.ers.usda.gov/data-products/major-land-uses/.

[6] T. Ulavi, T. Hebrink, J.H. Davidson, Analysis of a hybrid solar window for building integration, Sol. Energy. 105 (2014) 290–302. doi:10.1016/j.solener.2014.03.006.

[7] E. Gençer, C. Miskin, X. Sun, M.R. Khan, P. Bermel, M.A. Alam, R. Agrawal, Directing solar photons to sustainably meet food, energy, and water needs, Sci. Rep. (2017).

Copyright © American Institute of Chemical Engineers. All rights reserved.

498203 Embodied Phosphorus in Interstate U.S. Food Transfers: Sustainability Implications for Food-Energy-Water Nexus

Abstract:

Modern agriculture relies heavily on nitrogen, phosphorus, and potassium based fertilizers to reduce the yield gap and sustain current crop yields. Recently inorganic phosphorus fertilizers, an essential input to agriculture have garnered attention as they are primarily obtained from non-renewable and geographically concentrated rock phosphate sources. Phosphorus content in the soil is heavily skewed across the world with many regions facing water quality issues due to nutrient overloading in surface waters, while some face nutrient deficiency and soil degradation issues. Food trade is essential for supplying diverse and large quantities of food items to feed the growing population at global, national, and regional scale. Food trade also presents a pathway for transferring large volume of virtual nutrients and associated environmental impacts. The spatial heterogeneity of food production systems can increase or decrease the consumption based environmental impact of food trade in a particular region. At domestic level, vast quantities of food is traded between states resulting in a complex interstate food trade network. Although, estimates of United States (U.S.) centric domestic production and international trade of embodied phosphorus are available, we lack an understanding of interstate flux of embodied phosphorus for the U.S. To avoid unintended consequences and ensure optimal outcomes for a resilient food supply system, a holistic systems approach to studying interconnected resources is required. We model the origin and destination of food and embodied phosphorus flows to quantify trends in regional phosphorus use, identify hotspots, and propose interventions for enhancing the fertilizer use efficiency and overall environmental sustainability of the U.S. food system.

Using publicly available disparate datasets, we develop a food trade and associated virtual phosphorus model for the U.S. To this end, we evaluate over 60 crop items covering commodity groups of cereal grains, and fruits, nuts and vegetables. The interstate food trade network is translated to virtual phosphorus network by using crop yields and phosphate fertilizer application rates from estimates provided by the U.S. Department of Agriculture. Preliminary results indicate that over 655 million tons of food commodities were transferred across the U.S. in 2012 along with 61 million tons of embodied phosphorus fertilizer. Additionally, fruits and vegetables category contributed only 30% by mass for food trade but accounted for half of embodied phosphorus trade compared to cereal grains. We also evaluate virtual phosphorus savings for each individual item to identify phosphorus use efficiency in the US food production system. Additionally, we quantify the embodied energy flows associated with embodied phosphorus trade and discuss the inorganic phosphorus impact on water quality. Furthermore, we apply a network theory framework to understand the structure, robustness, and environmental sustainability of the considered trade networks. From a network perspective, our previous work has found that interstate food trade forms a well-connected, robust network with a majority of states participating in a high volume of trade. We characterize interstate virtual phosphorus network through series of higher order network measures to understand robustness and resilience patterns in virtual phosphorus trade. This will provide a necessary framework to evaluate series of interventions in a highly interconnected food trade system.

Copyright © American Institute of Chemical Engineers. All rights reserved.

495017 Comparing Beccs and DAC for Climate Change Mitigation: The Water-Land-Energy Nexus

Abstract:

The Paris Agreement signaled global commitment to limit average global temperature rise to 2°C and to pursue efforts to achieve 1.5°C. The IPCC AR5 cites negative emission technologies, particularly bio-energy with carbon capture and storage (BECCS), as critical to achieving this[1]. BECCS' ability to deliver negative emissions has been questioned with many uncertainties around land-use change emissions. Furthermore, large-scale biomass use poses several biophysical and socioeconomic risks such as exacerbating scarce water resources, biodiversity loss, and large opportunity costs of land-use for biomass. Capturing carbon dioxide (CO_2) directly from the atmosphere, *direct air capture* (DAC), presents an alternative source of negative emissions. Established capture technologies can be used to extract it from the atmosphere, seemingly without the significant natural resource requirements of BECCS. However, the low concentration of CO_2 in air necessitates a high energy cost of capture.

In this work, a comparison of the strengths and weaknesses of each technology based on key performance metrics (land, water and energy requirements) has been done. A negative emissions target of 12 Gt_{CO2}/y was used; this is the central estimate of what is required from RCP2.6[1] and has been cited as the target beyond which BECCS is no longer economically feasible. A *'farm-to-NET plant'* model of BECCS accounting extensively for the biomass supply chain and power plant specifications (coal co-firing proportions, capture rates) has been built to determine the overall GHG emission balance of the technology. Data from DAC processes proposed in the literature and existing pilot projects have been used to quantify the potential land, water and energy requirements of a large-scale system.

For BECCS, high variability in carbon balance outcomes was observed depending on the feedstock type and source, and assumptions on indirect land-use changes. Depending on decisions taken throughout the supply chain, a BECCS facility could take over 50 years to begin removing CO_2 from the atmosphere, or it could start removing CO_2 from day one[2]. Stark contrasts in resource requirements to meet the mitigation target were found for BECCS and DAC. Depending on the conditions of its deployment, a BECCS plant operating at 100% load factor could need 363-2100 Mha of land and 4-16 Tm^3/yr of water dedicated.

DAC's needs were 3-5 orders of magnitude lower at 0.04-3.3 Mha and 180 km^3/yr[3], thus posing little to these resources. It is important to note that DAC's water requirements are very sensitive to relative humidity and hence location of the plant[3]. Furthermore, the land area associated with the electricity used for DAC process was not accounted for, and could be significant when considering low carbon power, such as solar and wind.

BECCS' status as an energy-positive mitigation option was found to be untrue in some scenarios. Its annual energy output in the case of US-sourced Switchgrass could be as high as 22 EJ/yr, however up to 37 EJ/yr could be required if Brazil-sourced Miscanthus is the choice biomass. DAC energy costs, however, are still significantly higher at 81-274 EJ/y, even when the lower-bound from literature is assumed.

Surplus resource availability, particularly land and water, are highly-dependent on geography. BECCS and DAC have been shown to require contrasting amounts of these. Therefore, the preferred technology must be discussed within the context of a specific location, and in the case of BECCS, biomass source if not locally-grown.

References

1. IPCC. *Climate Change 2014: Mitigation of Climate Change. Working Group III Contribution to the Fifth Assessment Report of the Intergovernmental Panel on Climate Change* (2014). doi:10.1017/CBO9781107415416

2. Fajardy, M. & Mac Dowell, N. Can BECCS deliver sustainable and resource efficient negative emissions? *Energy Environ. Sci.* 0–42 (2017). doi:10.1039/C7EE00465F

3. Socolow, R. *et al.* Direct Air Capture of CO 2 with Chemicals Panel on Public Affairs. *Am. Phys. Soc. - Panel Public Aff.* 100 (2011).

Copyright © American Institute of Chemical Engineers. All rights reserved.

499908 Food Manufacturing: The Shift from Centralised to DistributedÂ Production

Abstract:

Projections for global food demand for a growing world population indicates an increase of 70% by 2050 (United Nations, 2017). The competitiveness among the food producers to satisfy production demand and consumer expectations has led to the search of new alternatives and innovations to make food products more attractive for the consumers, thus increasing company profits (Ingram et al., 2013).

With the growing concern about climate change, the consumer expectations are not limited to the relation quality-price, but also they include a low environmental impact of the product (Kemp et al., 2010).

As in other industries, the food manufacturing has benefited from the economy of scale, where the mass production allows the reduction of the manufacturing cost. However, a centralised production leads to a complex distribution chain, where the distances travelled for the product distribution increase and with them the cost and the CO_2emissions related to transportation.

Another production alternative is distributed manufacturing with shorter distances among the facilities (i.e. farm-factory, factory-distribution centre), thus the transportation cost and hence environmental impact are reduced. Nevertheless, other factors can affect the environmental impact of a production-distribution system, e.g. the carbon footprint of the raw materials.

The aim of this work is to analyse from an economic and environmental perspective several manufacturing-distribution scenarios, and identify the factors with greater influence in the choice between a centralised manufacturing and a more distributed one. For this purpose, a novel framework we call the *honeycomb model* is presented for the techno-economic and sustainability assessment of different manufacturing scenarios. This framework can be applied for the estimation of the optimum configuration that minimises the cost and CO_2emissions of the manufacturing-distribution chain, as well as perform studies of sensitivity of the optimum number of processing plants to certain parameters/characteristics of the system. Here, the tomato paste production is used as case study.

The results indicate that increments in the area of the system analysed and the product demand favour a more distributed manufacturing, while the origin of the raw materials and their carbon footprint could lead to the centralisation of the production.

Sensitivity analysis of the optimum manufacturing configuration provides important information about when a more distributed manufacturing is preferred over a centralised one. This information can be useful for further design of strategies to shift a centralised manufacturing-distribution configuration to a more distributed one when the purpose is to boost local economies.

This project is part of the Centre for Sustainable Energy Use in Food Chains (CSEF) from RCUK.

References

K. Kemp, A. Insch, D.K. Holdsworth, J.G. Knight. 2010. Food miles: Do UK consumers actually care? Food Policy 35:504â€'513.

J.S.I. Ingram, H.L. Wright, L. Foster, T. Aldred, D. Barling, T.G. Benton, et al. 2013.Priority research questions for the UK food system. Food Sec. 5:617â€"636.

United Nations. Available in: http://www.un.org/en/development/desa/news/population/2015-report.html. Date accessed: February 18th, 2017.

Copyright © American Institute of Chemical Engineers. All rights reserved.

494133 Understanding Biomass Value Chains and the Environment-Food-Energy-Water Nexus through Whole-Systems Analysis and Optimisation

Abstract:

Biomass has the potential to provide access to a sustainable supply of resources such as energy, chemicals and materials. Since the production of biomass, i.e. plant matter such as forestry, agricultural crops and residues, requires a large amount of land and water, competes with food production and can affect the ecosystem (e.g. deforestation, leaching of fertilisers and pesticides into water bodies), there is a delicate balance between all of these elements. Together, these are known as the environment-food-energy-water nexus and all of the benefits associated with the use of biomass to meet our requirements for energy and other products must be considered along with the potential impact to the nexus as a whole. Therefore, it is important not only to understand how best to exploit the great potential of biomass but also to understand the detailed interactions among all of the elements in the nexus.

The BEFEW (Biomass and the Energy-Food-Environment-Water Nexus) project funded by the Newton Fund and EPSRC (Grant No. EP/P018165/1) is examining the many different possibilities for producing energy and high-value products from biomass that also maintain the balance in the nexus. Detailed maps of existing land use and soil quality, water availability, quality and tolerance to contamination, and climate projections in terms of temperature and rainfall are used in order to understand the potential for growing biomass. Different processing facilities that convert the biomass to energy or other chemicals are researched and a comprehensive database of such technologies, including a site suitability analysis is developed. Thus, a detailed and rich representation of the influence, on the whole of the nexus, of the types of biomass grown, where they are grown, the types of technologies used (and therefore the products that will be made from biomass) and where they are located can be built. A mixed-integer linear programming (MILP) model is developed in order to determine the combinations of crops grown and technologies used to convert biomass to energy and high-value products (thus forming a biomass value chain) that provide the greatest benefit with the lowest impact on the nexus. Using the model, the efficient and robust value chains that are synergistic with the nexus and have the potential for deployment over a wide range of scenarios with different combinations of key socio-economic and environmental performance indicators and constraints are identified. In this conference, the latest progress in the BEFEW project will be presented, with a focus on the MILP model and case studies.

Copyright © American Institute of Chemical Engineers. All rights reserved.

505039 Innovative Approaches to Achieving FEW Nexus Goals that can Influence Near-Term & Long-Term Decision Making

Abstract:

The world population is estimated to reach to 9.7B by 2050, resulting in increased demand for resources and increased production of waste. As a result, today's waste disposal systems and approach to waste management will not be adequate for future applications. Therefore, a systematic approach to sustainable food, energy and water management is needed to meet increased demand for resources and efficient waste utilization systems. Currently, the food, energy and water (FEW) sectors use technologies developed to address the needs of a given sector (*i.e.* energy, water, food or agriculture), and wastes from each sector are usually managed separately. Production systems underlying these sectors have traditionally treated pollution and waste as externalities that are diluted into the ambient environment, and separate infrastructure systems have consequently been developed with unnecessary inefficiencies that are increasingly problematic. Because neither landfilling nor industrial composting provide ultimate reutilization of organic waste, and neither can achieve both energy and nutrient recovery in one process from organic waste, we need to transition from a "Traditional Linear Waste Disposal" approach to a sustainable "Closed Loop Resource Recovery" approach. In addition, finding efficient and affordable solutions to avoid GHG emission leakages from landfills and composting facilities through the use of Anaerobic Digestion infrastructure will be a key element of a sustainable system for future waste management. Thus, exploring how FEW systems can efficiently be optimized to (1) repurpose or recycle organic waste products, (2) internalize traditional externalities, and (3) integrate wastes with resource inputs across systems, by deploying integrative management strategies is essential. In addition, integrating other relevant disciplines that bring the necessary economic, social and educational perspectives to addressing this problem, will achieve broader impacts across the system.

Copyright © American Institute of Chemical Engineers. All rights reserved.

498677 Road Map for Embedding Ethics into ChE Undergraduate Curricula

Abstract:

The author will outline a number of methodologies and a suggested subject sequencing that can be effectively used to further develop an embedded ethics curricula. This curricula will consist of definitions, examples and suggestions on where to embed ethics into any existing ABET-approved, chemical engineering undergraduate course set. The author has lectured extensively at universities on ethics and a number of ethics related subjects.

Copyright © American Institute of Chemical Engineers. All rights reserved.

504018 Views on Ethics in Undergraduate Education

Abstract:

Engineering programs are expected to graduate students who understand their professional ethical responsibilities. A studentâ€™s experience with ethics often begins with academic integrity. Research on student attitudes and behavior regarding academic integrity will be reviewed with suggestions to help faculty nurture a strong ethical foundation in their students.

Copyright © American Institute of Chemical Engineers. All rights reserved.

495849 Ethical Reasoning in the Engineering Curriculum

Abstract:

Although the provision of ethics in engineering curricula has increased in recent years, the teaching of engineering ethics remains patchy and sporadic in a number of countries. Ethics in research, on the other hand, appears to be at a more mature stage. At the same time, the number of ethical codes, codes of practice and research concordats has flourished, both nationally and internationally. The questions remain: are those codes â€œliveâ€? Have those concordats resulted in an improvement of professional integrity? Is the teaching of ethics different and/or disconnected from ethics in research and in the workplace? Is ethics in chemical engineering different from ethics in other engineering disciplines? The paper analyses those questions in relation to an ever changing landscape, influenced by local and international drivers.

Copyright © American Institute of Chemical Engineers. All rights reserved.

RAPID

Transforming Process Industries

Breakthrough Innovation to Transform the Process Industries

Karen Fletcher
CEO - RAPID

Jim Bielenberg
CTO - RAPID

October 30, 2017

ManufacturingUSA

Manufacturing USA Overview

Vision

U.S. Global leadership in Advanced Manufacturing

Mission

Connecting people, ideas, and technology to solve industry-relevant advanced manufacturing challenges, thereby enhancing industrial competitiveness and economic growth, and strengthening our national security.

Goals

1. Increase competitiveness
2. Facilitate technology transition
3. Accelerate the manufacturing workforce
4. Ensure stable and sustainable infrastructure

www.ManufacturingUSA.com

2

RAPID
Transforming Process Industries

Manufacturing USA

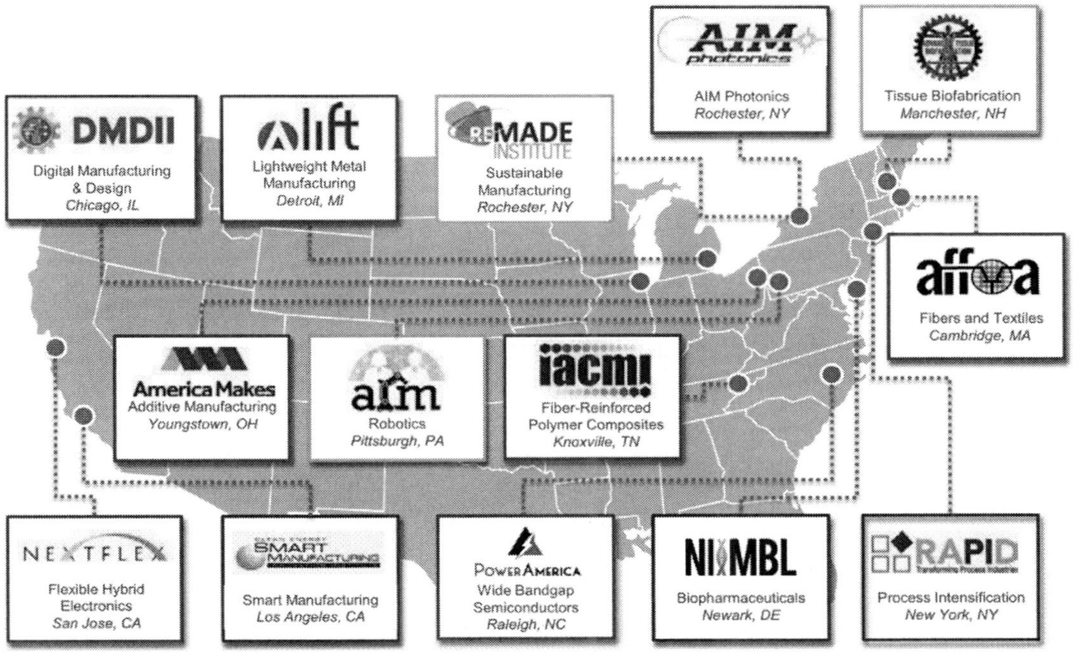

3

RAPID
Transforming Process Industries

RAPID's Industry-Led Vision

A dynamic network of partners who collectively build a sustainable ***ecosystem*** *that:*

… researches, develops and advances new technologies for modular chemical process intensification

… builds a strong portfolio of R&D projects & educational tools

… delivers dramatic reductions in energy, environmental footprint, capital and operating cost

… makes U.S. Manufacturing and our workforce more competitive

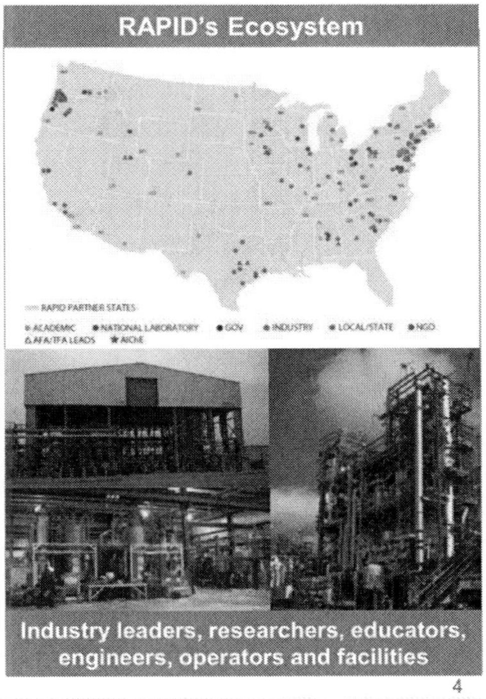

4

RAPID Transforming Process Industries

Our Mandate

- Lead a national effort to research, develop and demonstrate high-impact modular chemical process intensification solutions for U.S. Manufacturing.
- Actively build RAPID membership through an inclusive and attractive value proposition.
- Leverage $70 million of federal funding with cost share from members.
- Operate the Institute to benefit a wide range of stakeholders.
- Establish an infrastructure that enables access to process intensification resources, tools, expertise, and facilities.
- Bring together private and public entities to co-invest in R&D, commercialization, and deployment of innovative technologies.
- Establish a technical education and workforce development program.

"The goal for these Institutes is to revitalize American manufacturing and support domestic manufacturing competitiveness."
— U.S. DOE

RAPID Transforming Process Industries

DOE Performance Metrics

- Energy Efficiency
- Energy Productivity improvement
- Intensification in Individual Process Modules
- Cost-Effective Manufacturing of Modules
- Cost Effective Deployment
- Enabling R&D Portfolio
- Industrial Partnerships
- Pathway to Self-Sustainment
- Train the Trainers
- Educate Students
- Annual Planning Process
- Industrial Roadmap
- Emerging Supply Chain
- Diversity of Firms and Individuals in the Eco-System

6

RAPID Transforming Process Industries

52 Members and growing

RAPID Transforming Process Industries

Technical Focus Areas

Chemical Commodity Processes

- Develop guidelines for integration of novel reaction and separation modules
- Validate design tools for process intensification

Renewable Bio Products

- Prototype and scale novel bio-conversion processes
- Improve energy & capital efficiency of existing and emerging processes

Natural Gas Upgrading

- Maximize impact through transfer of technologies and learnings across industries
- Mature and demonstrate transformational and enabling technologies for gas utilization

8

RAPID Transforming Process Industries

Technical Focus Areas

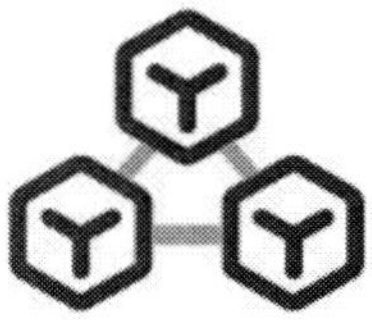

Modeling and Simulation

- Develop methods & tools for design, optimization, and intensification across multiple length and time scales
- Establish methods for intensification of dynamic/periodic operations

Intensified Process Fundamentals

- Advance inherently energy efficient separation processes & reaction platforms
- Develop fundamentals for multifunctional modules such as hybrid separation/ reaction schemes

Module Manufacturing

- Standardize modules and components to drive demand and capital investment within the supply chain
- Lower the cost of PI equipment using advanced manufacturing technology

9

RAPID Transforming Process Industries

2017 Accomplishments

- Completed DOE negotiations & all required documentation
- Stood up the Institute, with strong support/services from AIChE
- Established membership base
- Identified, negotiated and launched 4 Jump Start projects
- Formed Education & WFD Committee and completed a roadmap
- Completed focus area roadmaps with 100+ participants
- Executed call for pre-proposals and call for proposals
- Designed and launched a new website www.aiche.org/rapid
- Hired CTO (Mar), Technology Manager (Oct), Comms Leader (Nov)

10

RAPID
Transforming Process Industries

A Public-Private Partnership Between U.S. DOE EERE & AIChE®

Technology Gaps

RAPID Roadmap

Chemical & Commodity Processing

• Rxn/Separation
Novel reaction/separation schemes that are scalable and drive process efficiencies (e.g., membrane- or sorption-enhanced reactors). Applications include processing light paraffins to olefins, increase propylene yield vs conventional processes, hydrogen production, managing oxygen supply to reactions, etc.

• Non-Thermal Drivers
Use of alternative, non-thermal driving forces to activate chemical systems at the appropriate (atomic/molecular) scales.

• Batch Systems
Intensification schemes for batch systems. Transferring concepts largely developed for continuous processes to the batch realm could result in increased productivity/lower cost for specialty/fine chemicals.

Natural Gas Upgrading

• Selective Conversion
Concepts that show dramatic increases in desired product yields via fundamental improvements in catalysis, heat and mass transfer, and process concepts. This could include include alternative energy inputs and/or the use of novel reaction systems.

• Separations
Energy efficient separations technology to purify the reaction product mix, condition the feed in preparation for conversion, and to generate co-reactants for participation in natural gas conversion gas.

• Process Consolidation
Process consolidation and modularity to reduce total installed cost by reducing the total number of unit operations and by reducing the amount of field fabrication.

Renewable Bioproducts

• Primary Separation
Technologies to reduce energy demand in primary separation process steps designed to recover organic molecules and biomass components from water.

• Water management
Low capital and energy intensive solutions for dewatering and drying of biomass feedstocks, water removal and drying in pulp and paper process, and drying and removal of low levels of residual water from end products.

• Couple Rxn/Spn/Hxt
Use of novel chemistries and MCPI strategies to couple heat transfer and reaction in thermal processing of biomass and/or novel applications of reactive separation technologies in biological conversion technologies such as fermentation.

Intensified Process Fundamentals

• Scale-Out Methods
Scale out methodologies and models to predict performance of alternative energy input approaches for reactions and mixing and determine the suitable scale for modular manufacturing.

• Fundamental Data Acquisition/Modeling
Approaches to address key issues with lack of data on fluxes, adsorption, and catalyst kinetics for wide classes of materials, enabling model development and experimental testing of novel materials as adsorbents, membranes, catalysts and their integration.

• Predictive Models
Modeling capabilities to screen concepts and configurations of all types and predict optimal structures.

Modeling & Simulation

• PI Software Tools
Software tools for integrated reaction and/or separation processes and/or cyclic process such as pressure swing adsorption (PSA) or temperature swing adsorption (TSA). Such tools must be widely accessible and capable of integrating MCPI solutions with existing unit operations.

• Data Availability
Modeling approaches coupled with data generation and/or analysis to generate databases of physical parameters enabling design with mass separating agents.

• PI Assessment Tools
Tools to assess safety, sustainability, and control in PI and MCPI applications, including tools that address unique issues of uncertainty and reduced control variable options that are present in PI and MCPI applications.

Module Manufacturing

• Intensified Components
Intensified components that drive down the cost of module pre-assembly, transportation, and installation, while driving significant energy savings in chemical processes.

• Standard Designs
Design approaches that limit the amount of non-recurring engineering — during systems integration and installation — needed to support customized modules. This could include standard modules that enable economies of mass production and/or designs that enable incremental capacity additions.

• Distributed Processing
Module design and manufacturing approaches to enable distributed chemical processing. These will provide new paths to capital cost reduction and innovative techniques for maintenance and remote access and monitoring.

www.aiche.org/rapid

RAPID
Transforming Process Industries

A Public-Private Partnership Between U.S. DOE EERE & AIChE®

RAPID Education & Workforce Development Roadmap

OBJECTIVE: Establish a technical education and workforce development program that will leverage existing resources to train and educate the workforce, who can then research, develop, design and operate processes that incorporate new process intensification (PI) technologies and modular process designs widely within U.S. industry.

Undergraduate Students

GOAL
Create PI & MCPI awareness in future generations of engineers by augmenting the standard (ABET accredited) engineering curriculum with modules and educational content introducing PI & MCPI topics relevant to industry.

WHAT SUCCESS LOOKS LIKE
500 students* trained in PI & MCPI per year by 2019.

METHOD
eLearning course modules, face-to-face courses in conjunction with major meeting, Intern program.

Graduate Students

GOAL
Generate PI & MCPI expertise so that graduate students can connect the relevance of PI & MCPI research to industrial applications and create new opportunities and applications for MCPI.

WHAT SUCCESS LOOKS LIKE
500 students* trained in PI & MCPI per year by 2019.

METHOD
eLearning course modules and face-to-face courses in conjunction with major meeting.

Faculty

GOAL
Update and educate university faculty on PI & MCPI concepts and introduce eLearning modules created by RAPID for integration into their curriculum.

WHAT SUCCESS LOOKS LIKE
50 education/training professionals per year in MCPI by 2019.

METHOD
Face-to-face course in conjunction with major meeting on the Fundamentals of PI & MCPI.

Technicians/Operators

GOAL
Introduce technicians and operators to PI & MCPI, provide awareness of new processes that include PI & MCPI as well as emphasizing process safety.

WHAT SUCCESS LOOKS LIKE
At least 50 professionals** per year in MCPI by 2019.

METHOD
eLearning course modules.

Professional Engineers R&D | Design | Production

GOAL:
Train the current generation of engineers on how to utilize and deploy PI & MCPI technology in order to improve process efficiency.

WHAT SUCCESS LOOKS LIKE:
At least 50 professionals** per year in MCPI by 2019.

METHOD
eLearning course modules, face-to-face courses in conjunction with major meeting, and promotion through Manufacturing Extension Partnerships (MEPs).

Management

GOAL
Create high level awareness of the benefits of PI & MCPI from a techno-economic analysis perspective.

METHOD
Presentation material (includes relevant case studies, sustainability, and safety); short videos aimed at executives.

PI & MCPI Body Knowledge Drives Content Development Topics and Priority

Crosscutting areas identify content topics, levels of understanding for each audience member as well as high-priority topic areas. It includes four sub-sections:

- PI & MCPI Overview
- Modeling & Simulation
- Intensified Process Fundamentals
- Module Manufacturing

Application areas drive relevancy for content and provide industrial applications, examples and case studies within content and include the three sub-sections above.

* Students defined as undergraduate and graduate students.
** Professionals defined as technician, operator or engineer.

www.aiche.org/rapid

2017 Call for Proposals

RAPID Transforming Process Industries

- Overall Results
  - 73 submissions
  - 42 "lead" institutions
  - ~$30M (fed) request for 2018
- Proposals review process completed during Sept/Oct
- Announcement of selected projects November 2nd, 2017

# of Submissions	Submissions*
Chemical Commodity Processes	13
Intensified Process Fundamentals	13
Modeling and Simulation	6
Module Manufacturing	7
Natural Gas Upgrading	18
Renewable Bioproducts	16
	73

Summary Data - RAPID 2017 Proposal Call

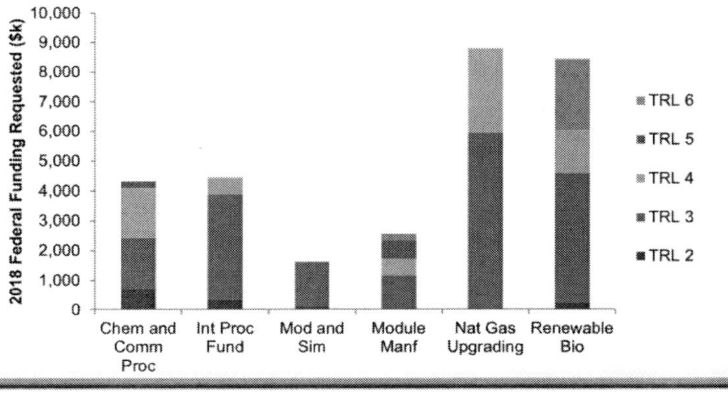

Dynamic Intensification of the Operation of Dividing Wall Columns

RAPID Transforming Process Industries

DOW

TEXAS The University of Texas at Austin

Innovation:

- Demonstrate that a dynamically controlled process (non-steady state) creates opportunity for improved efficiency

Project Outcomes

- Develop theoretical basis for periodic operation of intensified processes
- Experimental proof of concept using dividing wall column (DWC) pilot plant available at UT

Institute Target Addressed

- Develop tools and technologies to reduce the cost of deploying modular chemical process intensification in existing processes

Fall 2017 Update

- Theoretical framework for general value/screening of periodic operation tested
- Initial screening of DWC periodic operation assessed

RAPID Transforming Process Industries

Adsorptive Nitrogen Rejection from Natural Gas

PRAXAIR

Georgia Institute of Technology

Innovation

- Utilize breakthrough rate dependent adsorbent to modularize conventional PSA process

Project Outcomes

- Field demonstration of portable, modular PSA system

Institute Target Addressed

- 1,000 hours of pilot scale operating time for a modular and intensified process with reduced capacity , improved energy efficiency, and lower emissions/environmental waste

Fall 2017 Update

- System installation, shakedown, and testing to demonstrate adsorbent performance in a PSA pilot system underway
- Manufacturability studies progressing - potential adsorbent manufacturer identified.
- Early 2018 tasks - additional work on TEA and H_2S tolerance testing at Georgia Tech

15

RAPID Transforming Process Industries

Autothermal pyrolysis of lignocellulose wastes to sugars & other biobased products

easy energy SYSTEMS

IOWA STATE

Innovation

- Couple intensified reaction + heat transfer design and novel process chemistry to modularize sugar production from biomass

Project Outcomes

- Establish MCPI benefits of autothermal pyrolysis, leading to design of a commercial scale modular system

Institute Target Addressed

- MCPI RD&D resulting in a doubling of energy productivity ($ Returned/Energy Consumed) by a combination OPEX and CAPEX improvements.

Fall 2017 Update

- Demonstrate sugar yields in excess of BP1 targets on 1 kg/h pyrolysis reactor under auto thermal conditions
- Completed continuous pilot scale pretreatment of multiple feedstocks
- Continuing to work on longer time pilot scale testing

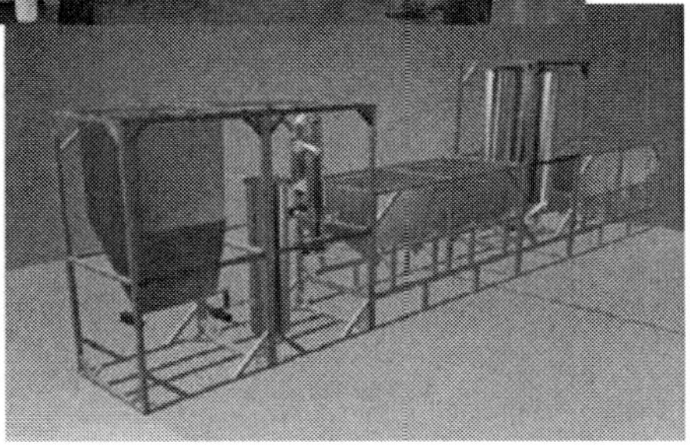

16

Mfg Supply Chain Development for the Modular Solar-Thermochemical Conversion Platform

RAPID Transforming Process Industries

Oregon State University OSU

Pacific Northwest National Laboratory

Innovation

- Coupling material processing approaches and manufacturing techniques that are uniquely well suited to small scales to generate low cost modular components

Project Outcomes

- Develop novel additive manufacturing techniques and materials that enable modular designs

Institute Target Addressed

- Scale-out manufacturing MCPI, with a unit cost reduction of >20% with each doubling in cumulative module manufacturing production

Fall 2017 Update

- OUS has developed unvalidated costs models for key manufacturing steps
- PNNL has established cost targets for key components
- OSU is working in-house capability to test new designs based on selective alloying
- Targeting 1Q2018 for new tool completion in support of novel manufacturing techniques

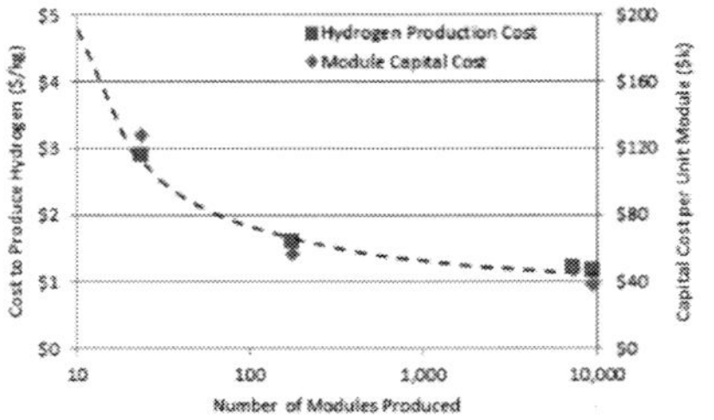

Estimated capital and product costs for individual STARS modules as a function of module mass production.

Plans for 2018

RAPID Transforming Process Industries

- Build the RAPID community
  - Member profiles, webinars, newsletters, website, networking opportunities
  - Conferences / face-to-face meetings (AIChE Spring / Annual)
- $30+M invested in R&D projects (fed funds + cost share)
- DOE (external) peer review
- Updated roadmaps; Gaps and new areas explored
- Project calls
- Training courses and educational webinars
- Increase membership and "reach"
  - New addition to our team: State MEPs*

* MEP = Manufacturing Extension Partnership

RAPID
Transforming Process Industries

To learn more about RAPID:

www.AICHE.org/RAPID

KareF@aiche.org

JameB@aiche.org

493251 Modeling and Simulation â€“ a Key Component in Enabling Process Intensification

Abstract:

The RAPID Manufacturing Institute aims to enable the development, de-risking, and implementation of process intensification across the chemical process industry. We will discuss critical needs in modeling and simulation in this area by highlighting the successes and challenges associated with using modeling and simulation to drive RAPIDâ€™s aims.

Copyright © American Institute of Chemical Engineers. All rights reserved.

493349 Modeling and Simulation Challenges for Process Intensification

Abstract:

Modeling and Simulation Challenges for Process Intensification

Efstratios N. Pistikopoulos[1,2], David Sholl[3], M. M. Faruque Hasan[1,2], Salih E. Demirel[1,2] and Yuhe Tian[1,2], (1)Artie McFerrin Department of Chemical Engineering, Texas A&M University, College Station, TX, (2)Texas A&M Energy Institute, Texas A&M University, College Station, TX, (3)School of Chemical & Biomolecular Engineering, Georgia Institute of Technology, Atlanta, GA

The recently established Rapid Advancement in Process Intensification Deployment (RAPID) Manufacturing Institute aims to transform the U.S. industry by developing the next generation of modular, intensified, and innovative manufacturing technologies and processes. The RAPID Modeling and Simulation Focus Area will in particular address the development of strategies and tools for the multi-scale representation, modeling, optimization, and control of intensified systems. In this presentation, we provide an overview of the challenges towards a systematic synthesis, modeling, and simulation approach for process intensification, as well as an exclusive review of process intensification synthesis approach methods in the open literature, such as the ones included in [1-5]. We also introduce an integrated framework for process intensification incorporating operability, control and safety (illustrated in Figure 1).

Figure 1: An Integrated Framework for Process Intensification Incorporating Operability, Control and Safety

[1] Demirel, S. E., Li, J., & Hasan, M. F. (2017). Systematic process intensification using building blocks. *Computers & Chemical Engineering*. http://doi.org/10.1016/j.compchemeng.2017.01.044.

[2] Baldea, M. (2015). From process integration to process intensification. *Computers & Chemical Engineering*, *81*, 104-114.

[3] Tula, A. K., Babi, D. K., Bottlaender, J., Eden, M. R., & Gani, R. (2017). A computer-aided software-tool for sustainable process synthesis-intensification. *Computers & Chemical Engineering*. http://doi.org/10.1016/j.compchemeng.2017.01.001

[4] Papalexandri, K. P., & Pistikopoulos, E. N. (1996). Generalized modular representation framework for process synthesis. *AIChE Journal*, *42*(4), 1010-1032.

[5] Ismail, S. R., Proios, P., & Pistikopoulos, E. N. (2001). Modular synthesis framework for combined separation/reaction systems. *AIChE Journal*, *47*(3), 629-649.

Copyright © American Institute of Chemical Engineers. All rights reserved.

494520 The Sustainable Synthesis-Design-Intensification of Chemical and Biochemical Processes

Abstract:

With the demands for efficient utilization of energy and water, better protection-control of the environment and correct-consistent use of raw materials, it is no longer sufficient to convert a selected raw material to a desired product, making profit and ensuring operability. The synthesized-designed process must now also be sustainable, incorporating not only economic feasibility but also environmental impact constraints, energy-water usage and/or waste constraints, resources availability constraints, product demand-price constraints and, many more. That is, the demands for new, innovative and sustainable processes need to be addressed. Process intensification has been identified as an important means to achieve this demand through integration with process synthesis-design in the early stage development of any chemical and biochemical process.

The presentation will highlight the issues related to defining the measures of sustainability through which new alternatives can be evaluated. Here, emphasis is given to the concept of sustainability, which is relative and therefore, a reference case first needs to be established. That is, compared to the base case, how sustainable are the new and innovative chemical and biochemical process designs?

A systematic multi-stage and multi-dimension computer aided framework which provides the necessary work-flow, the corresponding data-flow, the associated methods and the needed computational tools will be presented.

Through case studies involving chemical and biochemical processes, the application of the framework and the associated methods-tools to determine more sustainable solutions will be illustrated. It will be shown that significant improvements and non-tradeoff sustainable can be obtained through this integrated approach of combining process intensification as an essential element of process synthesis-design.

Copyright © American Institute of Chemical Engineers. All rights reserved.

2017 AIChE Annual Meeting

October 29 - November 3, 2017
Minneapolis Convention Center, Minneapolis, MN

Democratizing Energy Technology

Dane A. Boysen, PhD
Modular Chemical, Inc.

October 30, 2017

Overview

- How did RAPID get established?
- Why do we go big?
- Why is scale-up a problem?
- How do we break the hegemony of scale-up?

2011-2014

arpa·e

Advanced Research Projects Agency • ENERGY

2015-2016

gti®

GAS TECHNOLOGY INSTITUTE

Innovation Challenge

FINANCING ECONOMIES OF SCALE

GTL PLANT

(boe/d)	($/bpd)	plant cost
1	47,000k	$47M
10	10,000k	$100M
100	2,200k	$220M
1k	470k	$470M
10k	100k	$1,000M

AMMONIA PLANT

(ton/d)	($/tpd)	plant cost
1	10,000k	$10M
10	4,500	$25M
100	2,000k	$200M
1k	900k	$900M
10k	400k	$4,000M

calculated from "2/3 scaling law"

Innovation Challenge

FINANCING ECONOMIES OF SCALE

GTL PLANT

(boe/d)	($/bpd)	plant cost
1	47,000k	$47M
10	10,000k	$100M
100	2,200k	$220M
1k	470k	$470M
10k	100k	$1,000M

AMMONIA PLANT

(ton/d)	($/tpd)	plant cost	
1	10,000k	$10M	R&D demo
10	4,500	$25M	
100	2,000k	$200M	pilot demo
1k	900k	$900M	commercial
10k	400k	$4,000M	

calculated from "2/3 scaling law"

R&D Spending

OFFICE OF FOSSIL ENERGY BUDGET

U.S. DEPARTMENT OF ENERGY

FY 2015	$561M
FY 2014	$570M
FY 2013*	$495M
FY 2012*	$337M
FY 2011	$434M
FY 2010	$660M
FY 2009	$876M
FY 2008	$465M

Pilot chemical and power plants require **more than $200M**

How can we innovate with vastly inadequate funding?

*continuing resolution

FY2016 U.S. Senate Energy and Water Appropriations Bill

"The Committee recognizes that stranded, flared, and vented natural gas is the result of low natural gas prices that make transporting it uneconomic. As topics for additional Clean Energy Manufacturing Institutes are evaluated, the Secretary is encouraged to consider modular chemical processing as a way to address the issue of natural gas flaring and enable advanced manufacturing applications in the oil and gas industry."

https://www.congress.gov/congressional-report/114th-congress/senate-report/54/1

RAPID

U.S. DEPARTMENT OF ENERGY | Energy Efficiency & Renewable Energy

ADVANCED MANUFACTURING OFFICE

RAPID ADVANCEMENT IN PROCESS INTENSIFICATION DEPLOYMENT INSTITUTE

December 2016, the US Department of Energy announced the establishment of the 10th Manufacturing USA Institute, representing a critical step in the federal government's effort to double U.S. energy productivity by 2030. The Rapid Advancement in Process Intensification Deployment (RAPID) Institute is focused on addressing the barriers listed above to enable the development of breakthrough technologies to boost energy productivity and energy efficiency through manufacturing processes in industries such oil and gas, pulp and paper and various domestic chemical manufacturers. RAPID will leverage approaches to modular chemical process intensification (MCPI)—such as combining multiple process steps such as mixing, reaction, and separation into single more complex and intensified processes—with the **goal of improving productivity and efficiency, cutting operating costs, and reducing waste.**

https://www.aiche.org/rapid/about

RAPID Successes

1. Quick to get up and operational
2. Quality R&D projects in the que
3. Good industry engagement

Challenges for RAPID

1. No clear problem statement (goal ≠ problem)
2. No plan for establishing standards (modular = standards)
3. Not clear how progress will be tracked (need baseline)

RAPID Challenges

1. No clear problem statement (goal ≠ problem)
2. No plan for establishing standards (modular = standards)
3. Not clear how progress will be tracked (need baseline)

Why do we go big?

Energy Technology Today

PRICE TAG
$1-10 BN

COAL POWER PLANT

PRICE TAG
$0.5-5 BN

AMMONIA PLANT

PRICE TAG
$5-50 BN

GTL PLANT

Economies of Unit Scale

GAS-TO-LIQUIDS PLANTS

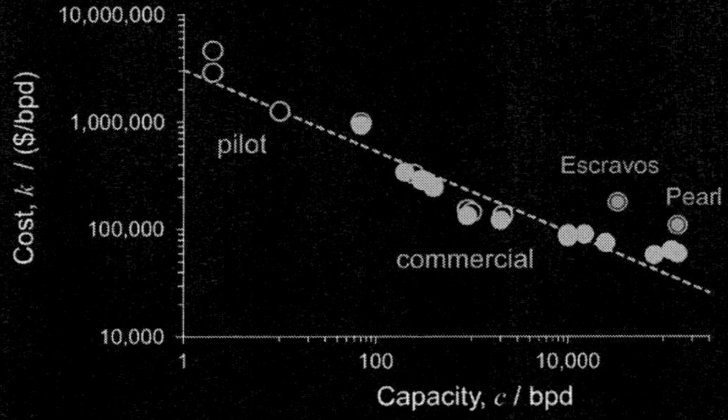

"2/3 scaling law"

$$k(c) = k(c_0)\left(\frac{c}{c_0}\right)^{\alpha}$$

Sources: (1) PJA Tijm. Gas to liquids, Fischer-Tropsch, advanced energy technology, future's pathway. Feb 2010; (2) C. Kopp. The US Air Force Synthetic Fuels Program. Technical Report APA-TR-2008-0102. (2008)

Why do we scale-up?

ARGUMENT

capital cost $\propto$ area [L^2]
capacity $\propto$ volume [L^3]
capital cost / capacity $\propto$ [L^2]/[L^3]

FLAW

pipe thickness ignored
hoop stress (constant) $\propto t / r$
capital cost $\propto l \cdot r^2$
capital cost / capacity $\propto$ constant

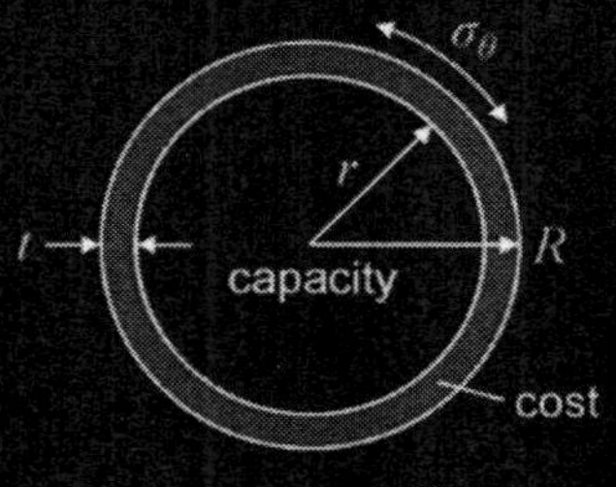

pipe cross-section

Why is scale-up a problem?

#1

BARRIER TO INNOVATION

#2

HUGE CAPITAL NEEDS

Shell Pearl GTL Facility, Qatar
MEGA-PROJECT ECONOMICS
RAND Study:
• 52 mega-projects
• $0.5B and $10B (1984 dollars)
• average over budget = 90%
Source: E.W. Merrow. Understanding the outcomes of megaprojects: a quantitative analysis of very large civilian projects. The RAND Corporation, Santa Monica, CA, 1988.

Capital Resources
U.S. COMPANIES BY REVENUE, 2007
10,000,000
1,000,000
100,000
10,000
1,000
100
10
1
Companies (number)
Revenues ($ billions)
8 companies
1,305,233
$61
3,302,803
$1,250
1,244,164
$3,474
176,850
$4,631
19,605
$11,004
775
$2,703
217
$5,169
8
$1,455
< $100k
< $1M
< $10M
< $100M
< $1B
< $10B
< $100B
> $100B
How many U.S. companies can finance a $1 billion plant?

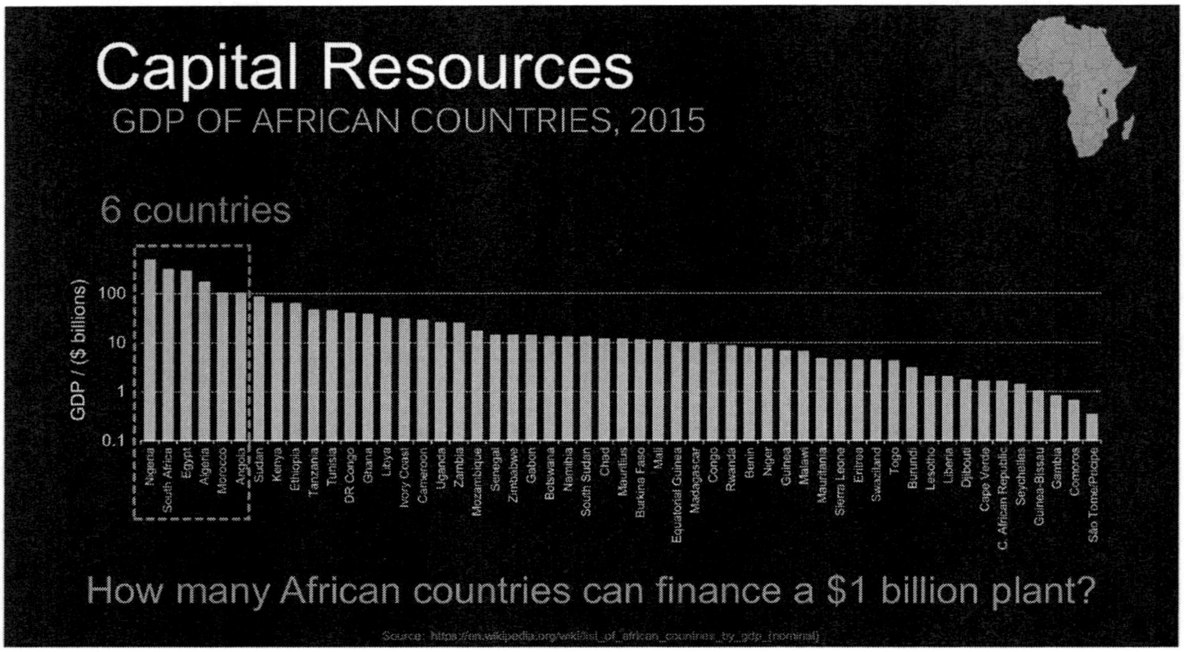
Capital Resources
GDP OF AFRICAN COUNTRIES, 2015
6 countries
GDP / ($ billions)
100
10
1
0.1
Nigeria
South Africa
Egypt
Algeria
Morocco
Angola
Sudan
Kenya
Ethiopia
Tanzania
Tunisia
DR Congo
Ghana
Libya
Ivory Coast
Cameroon
Uganda
Zambia
Mozambique
Senegal
Zimbabwe
Gabon
Botswana
Namibia
South Sudan
Chad
Mauritius
Burkina Faso
Mali
Equatorial Guinea
Madagascar
Congo
Rwanda
Benin
Niger
Guinea
Malawi
Mauritania
Sierra Leone
Eritrea
Swaziland
Togo
Burundi
Lesotho
Liberia
Djibouti
Cape Verde
C. African Republic
Seychelles
Guinea-Bissau
Gambia
Comoros
São Tomé/Pricipe
How many African countries can finance a $1 billion plant?

#3
DISTRIBUTED CHALLENGES

Developing World
DISTRIBUTED MARKETS

AFRICA
17% WORLD POPULATION
1% WORLD FERTILIZER USE

In 1999, Uganda farmers bought urea for $600/ton, global market price was $100/ton, why?

- Market size (< 1% global market)
- Transport cost (>$50/ton, 30% total)
- Finance cost ($300k, 1 kton)

Source: World Bank, 2015

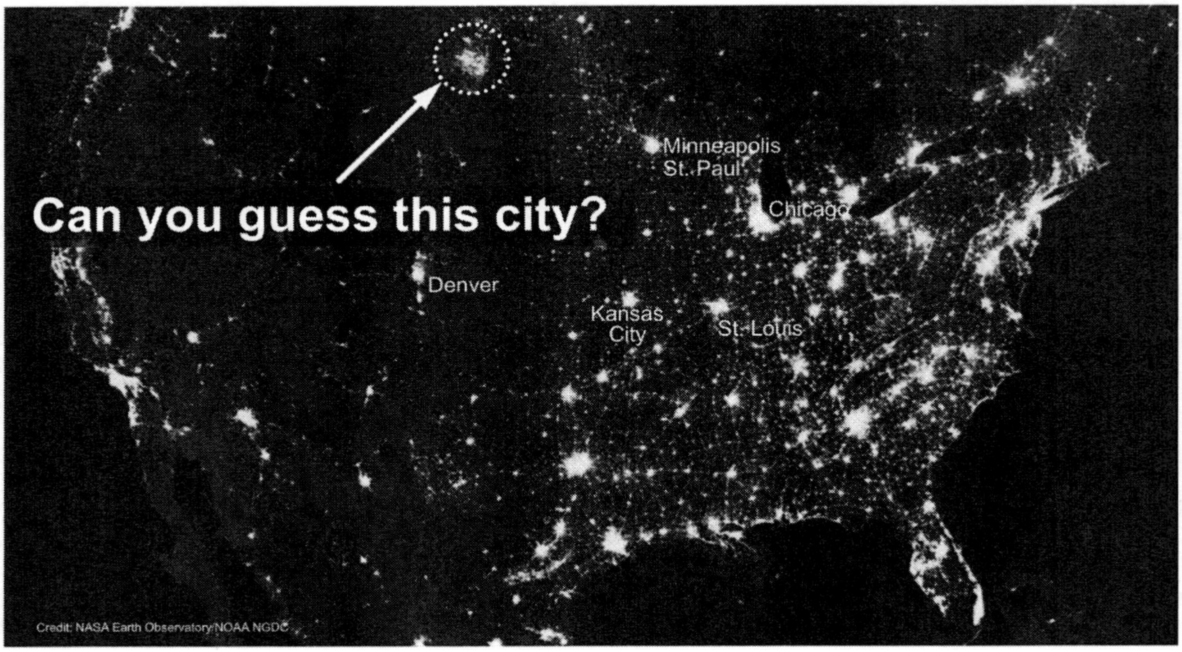
Can you guess this city?
Minneapolis
St. Paul
Chicago
Denver
Kansas City
St. Louis
Credit: NASA Earth Observatory/NOAA NGDC

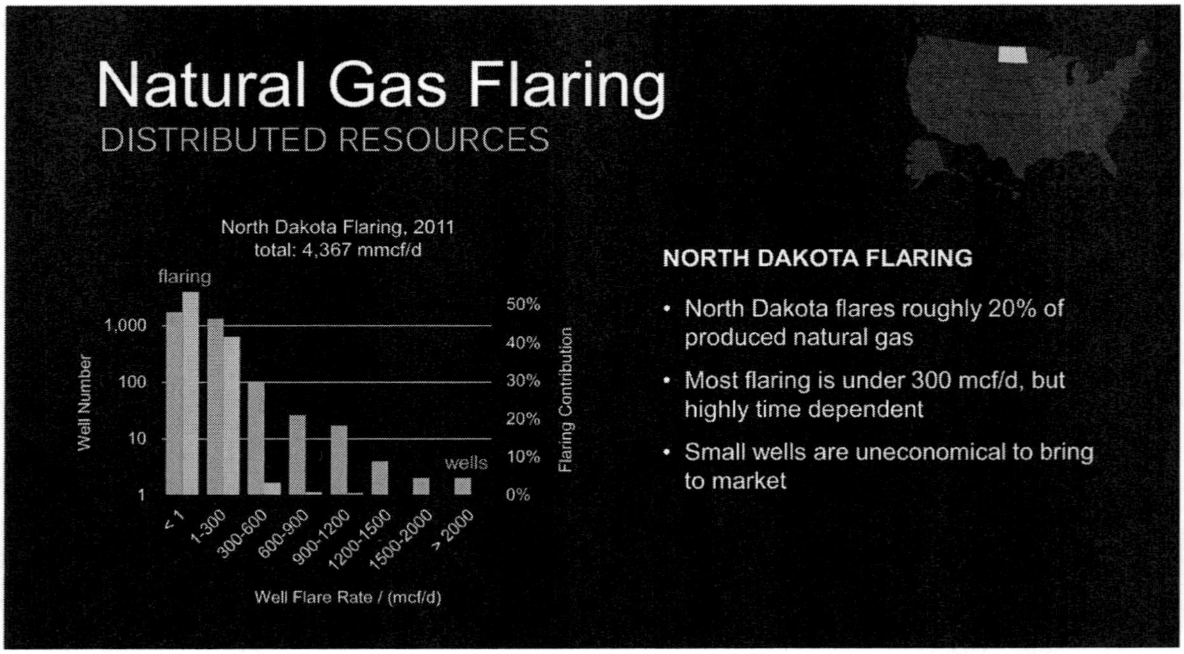
Natural Gas Flaring
DISTRIBUTED RESOURCES
North Dakota Flaring, 2011
total: 4,367 mmcf/d
flaring
wells
Well Number
1,000
100
10
1
50%
40%
30%
20%
10%
0%
Flaring Contribution
<1
1-300
300-600
600-900
900-1200
1200-1500
1500-2000
> 2000
Well Flare Rate / (mcf/d)
NORTH DAKOTA FLARING
• North Dakota flares roughly 20% of produced natural gas
• Most flaring is under 300 mcf/d, but highly time dependent
• Small wells are uneconomical to bring to market

How do we break the hegemony of scale-up?

Democratizing Technology

HISTORIC EXAMPLES

1450	1908	1950	1973	1977
Gütenburg Press	Ford Model T	ISO Shipping Container	Motorola DynaTAC 8000X	Commodore PET
INFORMATION	TRANSPORTATION	SHIPPING	COMMUNICATION	COMPUTATION

Democratizing Technology

WHAT ARE THE DEFINING FEATURES?

1. Empowers the little guy
2. Never developed by the incumbent
3. Displaces entrenched incumbent
4. Levels the playing field
5. Leverages capital of the many
6. Modular, mass produced, standardized
7. Fast innovation cycles

...and there are many examples

guns	*...displaced*	*swords*
Gutenberg press	*...displaced*	*illuminated manuscripts*
iso-containers	*...displaced*	*bulk shipping*
cellular phones	*...displaced*	*land lines*
personal computers	*...displaced*	*central computing*
automobiles	*...displaced*	*horses*
photovoltaics	*...displaced*	*solar thermal*
electric-arc furnaces	*...displaced*	*blast furnaces*
e-commerce	*...displaced*	*store fronts*
televisions	*...displaced*	*theatres*
record players	*...displaced*	*live musicians*

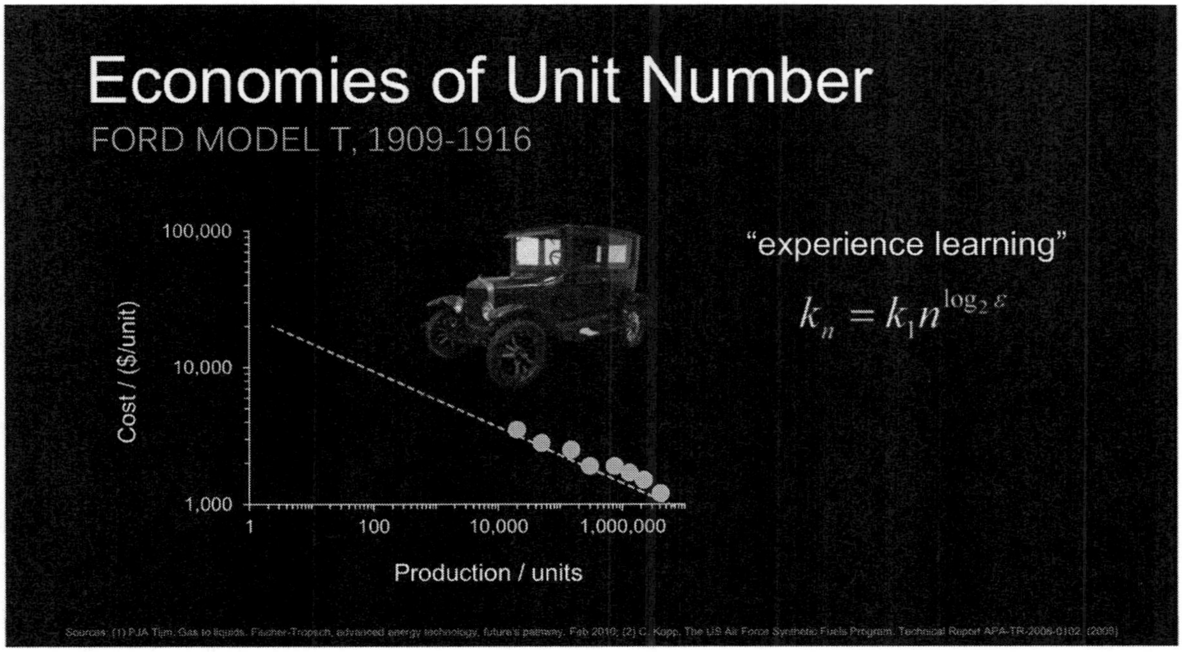

Economies of Unit Number
FORD MODEL T, 1909-1916
100,000
10,000
1,000
Cost / ($/unit)
1
100
10,000
1,000,000
Production / units
"experience learning"
$k_n = k_1 n^{\log_2 \varepsilon}$
Sources: (1) PJA Tijm, Gas to liquids, Fischer-Tropsch, advanced energy technology, future's pathway, Feb 2010; (2) C. Kopp, The US Air Force Synthetic Fuels Program, Technical Report APA-TR-2008-0102 (2008)

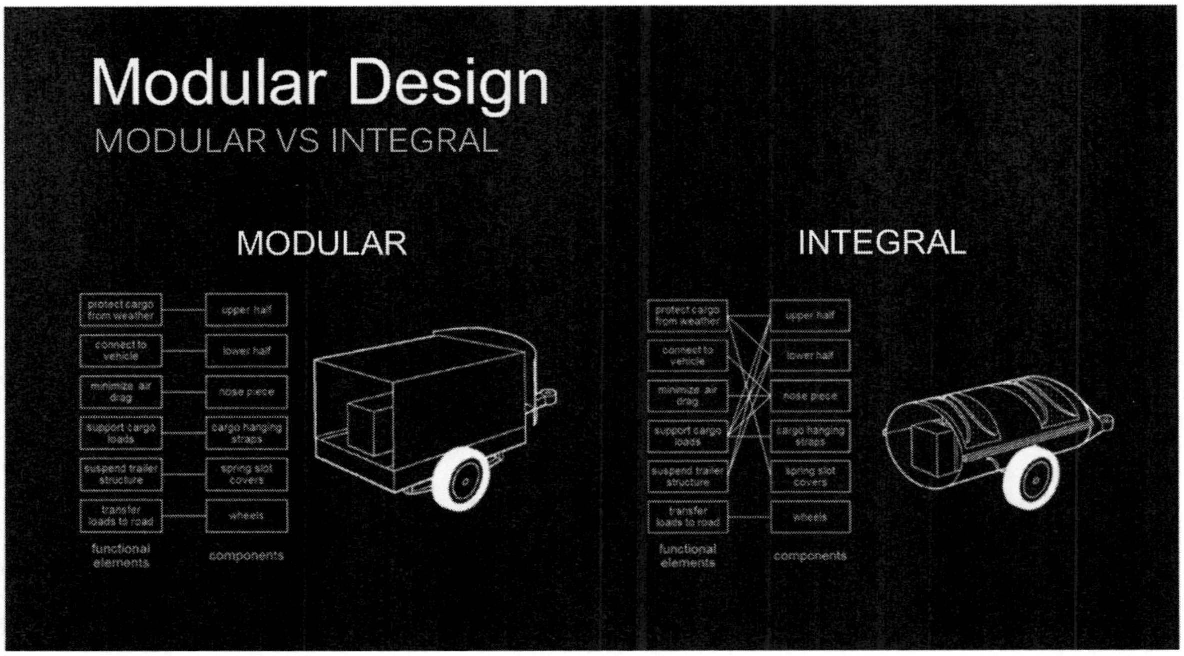

Modular Design
MODULAR VS INTEGRAL
MODULAR
INTEGRAL
protect cargo from weather
connect to vehicle
minimize air drag
support cargo loads
suspend trailer structure
transfer loads to road
functional elements
upper half
lower half
nose piece
cargo hanging straps
spring slot covers
wheels
components

Modular Design

MODULAR VS INTEGRAL

MODULAR

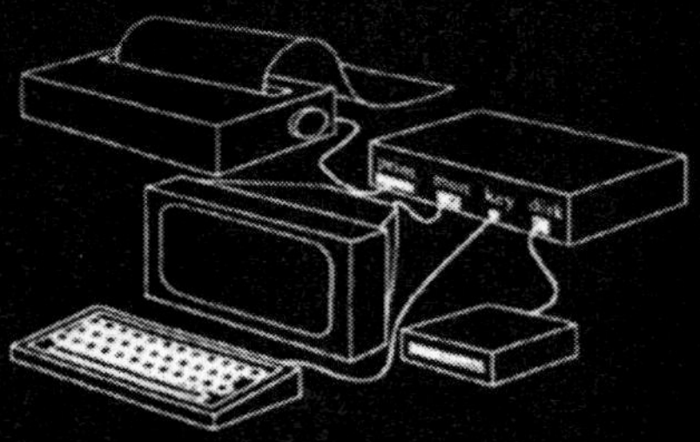

INTEGRAL

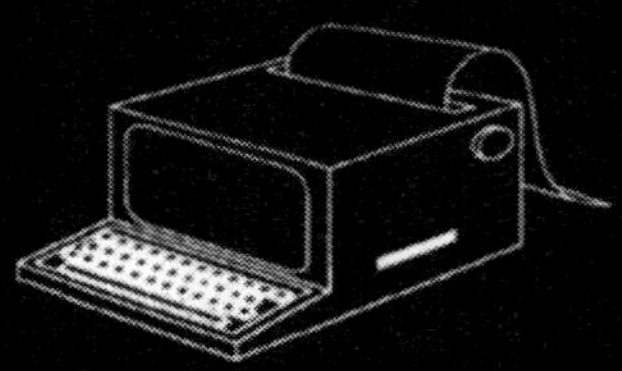

Small Modular

OIL REFINERY (WORLD)

- 7 TW petroleum refining
- 700 plants
- $500/kW capex

AUTO ENGINES (U.S.)

- 1.3 TW motive power
- 250 million engines
- $50/kW capex

Small Modular

GAS TURBINES (U.S.)

- 0.2 TW electricity
- 5000 gas turbine generators
- $1000/kW capex

AUTO ENGINES (U.S.)

- 1.3 TW motive power
- 250 million engines
- $50/kW capex

What about the democratization of chemical processing?

not if, but
when, who, how

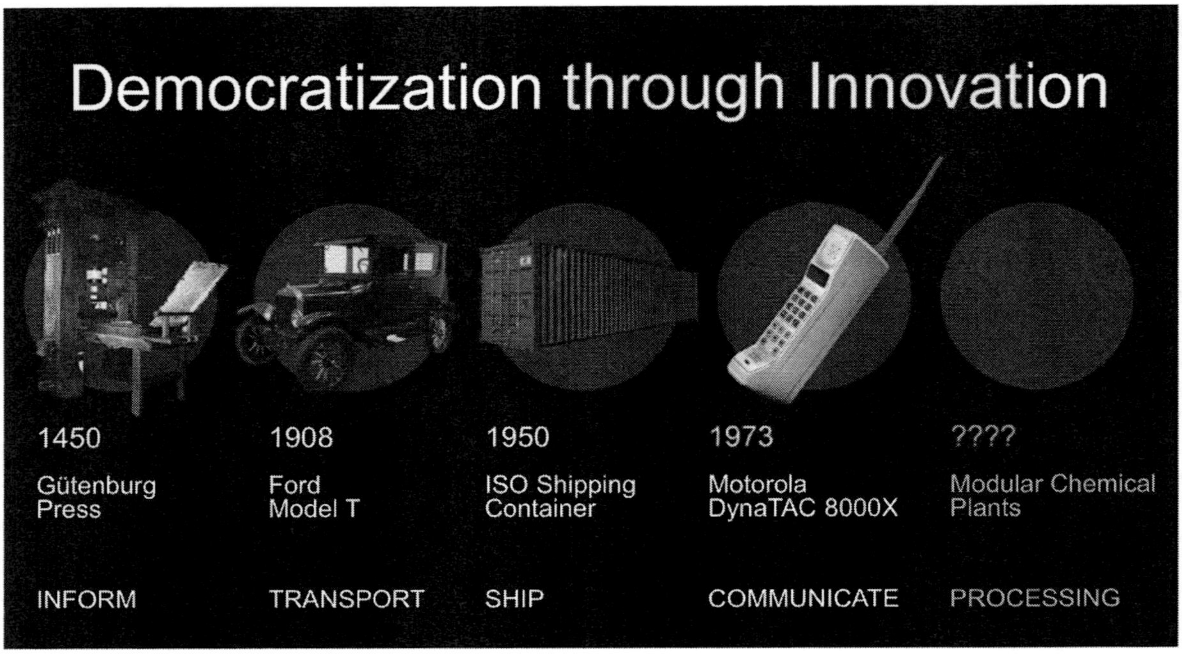

Democratization through Innovation
1450
Gütenburg
Press
INFORM
1908
Ford
Model T
TRANSPORT
1950
ISO Shipping
Container
SHIP
1973
Motorola
DynaTAC 8000X
COMMUNICATE
????
Modular Chemical
Plants
PROCESSING

Isaac Mkalia, 20, checks his mobile phone in Kojiado district, near the
(Photograph: Sven Torfinn/Oxfam)

Thank You

Dane A. Boysen
dane.boysen@gmail.com

494685 Evaluating Combined Heat and Power Deionization Systems for Efficient Water Reuse at Thermoelectric Power Plants

Abstract:

Efforts to increase both the United States and global water supply has primarily taken place through seawater and brackish water desalination. According to the International Desalination Association, the United States capacity for desalination increased in 2013 by 50% from 4 million to 6 million m^3/day. Unlike most areas of the world, nearly 77% of all desalination in the United States comes from the treatment of brackish water. Brackish water has lower salt concentrations (~<10 g/l) than seawater (35 g/l) and thus the energy consumed to remove salt ions in theory is much less than seawater desalination. Current technologies utilized for seawater desalination, reverse osmosis (RO) and Membrane distillation (MD), require a significant amount of energy regardless of the type of water (brackish or seawater). Furthermore, these technologies have approached their thermodynamic energy limits (1-3 kWh $m^{â€“3}$), indicating that further energy savings cannot be achieved. RO and MD are energy intensive technologies because they operated through extracting the main component water (H_2O) from the salt (NaCl). Alternatively, capacitive deionization (CDI), an emerging separation-based technology, operates by adsorbing the minority component (NaCl) from the feeding brackish water. Traditional CDI cells operate in a cyclic process where the capacitive electrodes are brought in contact with a concentrated salt solution (e.g. brackish water) and charged. Once the charged electrodes are fully saturated with salt ions, the electrodes are regenerated through discharging the electrodes in a brine solution. Analogous to a P-V diagram for traditional heat engines, the enclosed area of the charge versus voltage represents the energy that is consumed during this charging-discharging process. Until recently the CDI three step intermittent process has been relatively untouched in terms of literature and system operation. In this study, we explore the effect of temperature on salt removal with CDI. Experimentally, we use traditional electrochemical characterization methods, cyclic voltammetry (CV) and electrochemical impedance spectroscopy (EIS), to evaluate the capacitance and resistance changes with temperature. Here it was found that through increasing the operating temperature from 25°C to 45°C, the specific capacitance of carbon electrodes increased from 34.60±2.15 to 42.56±3.57 F/g, and the cell resistance decreased from 20.20±2.97 to 15.57±2.52 ohm. Additionally, the salt adsorption capacity decreased from 13.82±1.13, to 6.58±1.01 milligram NaCl/mg carbon. This indicates that although the electrochemical performance of CDI can be increased with temperature, the ion storage within the electric double layer is hindered at higher temperatures.

Copyright © American Institute of Chemical Engineers. All rights reserved.

493557 Biochar Amendments for Increased Crop Yields: How Can Biochars Improve Crop Nutrient Availability?

Abstract:

Background:

Maintaining a sustainable food supply for the foreseeable future without excessive environmental degradation will require novel engineering approaches for remediating the human disruption of the nitrogen cycle. Over the past several decades, food production has been able to keep pace with human population growth only thanks to the development of new high-yielding crop varieties grown with the help of fertilizers. Thus, vast amounts of energy are consumed every year to produce the approximately 200 Mt of ammonia needed to make the nitrogen (N) fertilizers used by farmers around the globe.

Unfortunately, more than half of the N fertilizer applied to a field is not available for plant growth due to losses caused by surface runoff, leaching into surface and ground water, or volatilization. Because of these losses, the increased use of nitrogen fertilizers has been linked to a variety of water pollution problems ranging from the expansion of the hypoxic zone in the Gulf of Mexico caused by eutrophication to the contamination of wells and groundwater with nitrate N. Moreover, denitrification reactions convert nitrates to nitrous oxide, a major greenhouse gas and nitrogen oxide, a major atmospheric pollutant and significant contributor to the depletion of atmospheric ozone.

The amendment of soils with biochar has been heralded as a sustainable method for increasing crop yield and preventing pollution problems caused by fertilizer runoff. However, the available experimental results are highly variable [1-3]. While some studies reported significant beneficial effects with up to 60% higher yield after biochar addition, other studies reported that biochar amendments either had no effect or even resulted in up to 30% lower agricultural productivity [3]. One of the most important reasons for this variability is that biochar is not a single entity. Dozens of biomass feedstocks are used to produce biochars in multiple types of reactors under varying temperature and oxygen conditions, leading to thousands of biochars with widely varying properties. To complicate things even more, soil properties, climate conditions (like rainfall and temperature), plant requirements and many other parameters vary from application to application. As a result, such studies and meta-analyses have not provided much insight into the fundamental mechanisms that modulate the environmental performance of biochar amendments.

Our Previous Work on this Problem:

To begin bridging this knowledge gap, we focused our recent research on the mechanisms that control the ability of biochars to improve crop nutrient availability, one of the main mechanisms proposed to explain crop yield increases in biochar-amended soils [2].

As a first step in this direction, we have recently developed and tested a mathematical model that uses first principles to describe nutrient transport in soils amended with biochar [4]. Using a simple Langmuir isotherm to describe nutrient adsorption, our simulation results revealed that nutrient transport in biochar-amended soils is modulated by a complex interplay between the physical mechanisms governing flow in porous media and nutrient transport, and the chemical mechanisms modulating nutrient adsorption of the pore surface of biochar solids. Nutrient transport rates through the topsoil decreased when biochars with the proper adsorption capacity and affinity were mixed with the soil. This was due to the differential dynamics of the adsorption and desorption processes. Nutrient adsorption was significantly faster than desorption, with biochar essentially acting as a slow (or controlled) release medium for the adsorbed nutrient [4].

Current Work and Summary of Results:

The present study extends our previous work with a systematic study of the adsorption and desorption dynamics on a much broader class of biochars. We report the development of a mathematical model that describes the response of biochar particles to a pulse of nutrient that simulates a single fertilizer application. Nutrient adsorbs on the biochar particle when the fertilizer solution flows past the particle and starts desorbing when the fertilizer solution is replaced by pure water. Two nutrient transport mechanisms are considered: (a) diffusion through liquid-filled pores and (b) surface diffusion. We assume that nutrient adsorption follows a Sips isotherm whose parameters can be varied to describe adsorption on both homogeneous and heterogeneous solids. The Freundlich and Langmuir isotherms commonly used to describe adsorption of ammonium or nitrates on biochars are special cases of the Sips isotherm. The highly nonlinear partial differential equations describing this transient adsorption/desorption problem are discretized using high-order finite differences and the resulting system of first order differential equations are integrated with a variable-step, variable-order integrator based on numerical differentiation formulas.

As noted previously, a "good" biochar should be able to quickly adsorb the nutrient when it is present in the liquid flowing around it, but it should desorb it much more slowly in order to act as slow (or controlled) release medium. This is the mechanism through which a "good" biochar will improve crop nutrient availability [2, 4].

Our simulations clearly show that the ability of a biochar to slowly release the adsorbed nutrient depends on a complex interaction of (a) external mass transfer, (b) intraparticle diffusion (both pore and surface diffusion), and (c) adsorption dynamics (affinity constant and adsorption capacity).

The interaction between external mass transfer and intraparticle pore diffusion is a key factor for modulating the release of nutrient that has adsorbed on the biochar. Our system is characterized by low values of the Biot number, the dimensionless number that represents the ratio of external mass transport rate over the intraparticle diffusion rate. This is because of the low interstitial water velocities (10-90 cm/day) in biochar-amended soils. Low Biot numbers are one of the key factors responsible for the long desorption tails that characterize the slow release of nutrients from "good" biochars.

Our simulations reveal that all the processes mentioned above can significantly affect the nutrient release rates. For example, surface diffusion accelerates nutrient desorption from small (2 mm in diameter) and highly porous biochar particles (0.6 total porosity). But, surface diffusion will slow down desorption from larger (4-8 mm in diameter) and less porous particles, even though the latter biochar particles may have exactly the same adsorption isotherm parameters. Adsorption affinity and site heterogeneity can also interact with porosity and particle size to modulate nutrient desorption. Finally, surface diffusion again improves the ability of a biochar to act a slow release medium at high interstitial velocities.

By revealing that the response of biochar particle to a pulse of nutrient is governed by complex interactions among external mass transfer, intraparticle diffusion and adsorption, this study provides new insights into the fundamental mechanisms that influence the environmental performance of biochar amendments. It may also explain why biochars that share some key properties (e.g. carbon content) or produced at the same pyrolysis temperature can potentially have widely different impacts on crop yields, as has been repeatedly reported in the literature [2,3]. Much better characterization of biochar properties (porosity, adsorption isotherm, intraparticle transport mechanism) and more precise control of amendment or soil parameters (like particle size, biochar aging, soil permeability) are necessary before we can fully elucidate the mechanisms of biochar action.

References:

[1] Spokas, K. A.; Cantrell, K. B.; Novak, J. M.; et al. "Biochar: A Synthesis of Its Agronomic Impact beyond Carbon Sequestration." *J. Environmental Quality* **41** (2012) 973-989.

[2] Jeffery, S; F.G.A. Verheijena; M. van der Velde; A.C. Bastos. "A quantitative review of the effects of biochar application to soils on crop productivity using meta-analysis." *Agriculture, Ecosystems and Environment* **144**(2011) 175–187.

[3] Crane-Droesch, A; S. Abiven; S. Jeffery; M.S. Torn. "Heterogeneous global crop yield response to biochar: a meta-regression analysis." *Environmental Research Letters* **8**(2013) 044049.

[4] Sun, S; C. E. Brewer; C. A. Masiello; K. Zygourakis. "Nutrient Transport in Soils Amended with Biochar: A Transient Model with Two Stationary Phases and Intraparticle Diffusion" *Industrial and Engineering Chemistry Research*, **54** (2015) 4123–4135.

Copyright © American Institute of Chemical Engineers. All rights reserved.

501354 Moisture Retention in Emulated Soil Micromodels: Development and Performance of Sustainable Agriculture Biotechnology

Abstract:

Soil bacteria living near plant roots secrete exopolysaccharides (EPS), a polymeric substance which can help modulate soil moisture and promotes plant growth. EPS functions in several different ways: (i) EPS behave as hydrogel, swelling during wet soil conditions and remaining hydrated during dry soil conditions; (ii) EPS promotes aggregation of soil particles and thereby promotes water retention in dry soil conditions through stronger capillary forces; and (iii) EPS alters surface properties of soil, creating more water repellent surfaces to inhibit rapid re-wetting in wet soil conditions. EPS functions in concert with the physical microstructure of real soil; therefore, it is necessary to study EPS performance in physically realistic micro-structures. Here we show pore-scale changes in moisture distribution within emulated soil micromodels that have systematically varied physical microstructures and surface water repellencies, and can be loaded with well-defined pore solutions. Experimental results showed dilute EPS solutions took twice as long to dry compared with pure water in identical micromodels. Microstructure geometry was found to influence drying rate over a range of pore saturation values as well as control the spatial distribution of liquid and gas phases within unsaturated soil micromodels. Deionized water took four times longer to completely dry from micromodels with higher water repellency surfaces compared with micromodels with lower water repellency surfaces. These results illustrate the various mechanisms whereby EPS impacts the spatial distribution and evaporation resistance of soil moisture. We provide a systematic method to evaluate performance of natural EPS solutions in modulating soil water and promoting resiliency of terrestrial ecosystem; we anticipate our approach may speed the development of sustainable agriculture biotechnology to enable food production using less water.

Copyright © American Institute of Chemical Engineers. All rights reserved.

498287 Molecular Modeling and Simulation Studies of the Structural and Energetic Evolution during Dehydration of Food Systems

Abstract:

From a molecular point of view, foodstuffs are mostly comprised of biopolymeric molecules that form highly porous structures in which solvents, solutes, additives, and other natural or artificial ingredients are accommodated. Because of the complexity of the food structures and compositions and because of the sensitivity of the pore structures and properties of food to physical and chemical changes arising externally or internally, foodstuffs are one of the most complex types of soft materials and are inclined to exhibit significant structure-property relationships in the physicochemical, functional, processing, and nutritional aspects of food science and engineering. In this respect, water (moisture) is a major factor in determining the properties of foods in various stages as it is not only a major component of food materials but also acts as space filler for stabilizing food structures, solvent for dissolving additive molecules, medium for heat and mass transfer, and participant in the chemical and/or biological reactions in food systems. .

One of the important ways for preparing and preserving foods is dehydration (water removal), which is a process that requires energy to be supplied to enable water molecules to break loose from their interactions and escape from the porous structures of the foods. The process of food dehydration and the preservation of dehydrated foods can thus be directly related to the interactions of water molecules with food macromolecules, solute ions, and additive molecules, because these interactions dictate the energy requirement for water removal during food dehydration as well as the structures of foods before, during, and after dehydration. Moreover, the magnitudes of these interactions depend on the densities and intermolecular distances of mutually interacting molecules, rendering the water molecules in foods to exist in different energetic states and making single system-wide thermodynamic parameters (e.g, water activity) inadequate in representing the diverse states of water molecules in food porous structures and in elucidating the microscopic and macroscopic structural evolution of food materials caused by dehydration. Capable of tackling these challenges is the molecular dynamics (MD) method which employs molecular-based interaction potentials to construct a model system that provides a satisfactory physicochemical representation of the real system and allows various structural and dynamic properties to be studied at a molecular level. Recently, we have employed MD modeling and simulations to construct an amylose-based food material and study the dehydration process in a continuous dynamic manner by periodically examining the energetics of all water molecules at the water-vapor interface and removing the one with the highest energy from the model food system. It was found that the strong interactions of water molecules with the food polymer stabilize the porous structure of the food material and makes the dehydration energy requirement to be greater than that involved in pure water vaporization. They also cause the water interaction energetics to be highly nonuniform in all three spatial dimensions. As dehydration proceeds, the surface tension of the descending water interface coupled with strong water-polymer interactions exert downward stresses on the food polymer to induce polymer conformational compaction and food structural reduction.

We further added two functional food ingredients to the model food systems in our MD simulation studies. The first one is potassium sorbate, which is a commonly used food preservative against the growth of bacteria, fungi, molds, and yeast in foods. The second one is fructose-1,6-diphosphate (FDP), which is a bioprotective agent and a promptly available high energy substrate and can promote growth and improve health. The stability of the food porous structures was found to be enhanced by the presence of the food additive molecules which function as additional fillers in the food pore space and interact strongly with the polymer chains and water molecules of the food material.Â During dehydration, the descending water level tends to accumulate the additive molecules, in particular sorbate with a nonpolar hydrocarbon moiety, at the vapor-liquid interface, which helps to form dense clusters of food polymer chains at the interface, reduces the level of structural reduction, but increases the energy requirement for water removal.Â The additive molecules in the interior regions of the food structures have strong tendencies to stay close to food polymer and hence exhibit nonuniform density distributions as well. Extensively dehydrated food porous structures were found to have substantially improved stability due to the water molecules remaining in the pore structure that effectively â €œbindâ€ the food polymer and additive molecules together. Which makes it very difficult to recover the original pore structures of the food systems that existed before the commencement of dehydration. The findings of these MD studies suggest that the original pore structures could be reduced less during dehydration and could be recovered more effectively by rehydration, if (i) the food material has a higher polymer density, (ii) solute or additive molecules have elongated shapes to act as cross-linkers, and (iii) the effects of water surface tension can be minimized or eliminated using special dehydration methods such as freeze drying. The MD modeling and simulation approach developed in this work could form a basis for systematic evaluation of food porous structures and their structural and energetic evolution and mechanisms in molecular detail and for providing scientific explanations to the macroscopic trends and empirically observed phenomena during food dehydration.

Copyright © American Institute of Chemical Engineers. All rights reserved.

493109 Metallic Membranes for N_2 Separation and NH_3 Production

Abstract:

The availability of nitrogen, phosphorus, and water are the three main factors that limit our ability to produce enough food to feed the growing population of the planet. The nitrogen cycle is one of the most significant biogeochemical cycles on Earth, as nitrogen is an essential nutrient for all forms of life. Although freely available in the atmosphere as dinitrogen (N_2), access to fixed forms of nitrogen constitutes, in many cases, the most limiting factor for plant growth. The industrial production of ammonia for fertilizers via the current Haber-Bosch process is an energy-intensive process that consumes 1-2% of the worldâ€™s annual energy supply. For these reasons, the need for advanced catalytic methods for the reduction of N_2 to ammonia remains a requirement for sustainability in the food, energy and water systems cycle.

The aim of this work is to explore the potential of metallic membranes for N_2 separation with the final intent to produce NH_3. Based on a preliminary theoretical investigation using density functional theory, the Group V transition metals (e.g., vanadium (V), niobium (Nb) and tantalum (Ta)) show strong affinity toward N_2. Moreover, from solubility and diffusivity values taken from the literature, iron (Fe) is a suitable fit for this application. Therefore, V, Nb, Ta and Fe foils with a thickness of 40 Î¼m have been chosen as membrane materials in this study.

Permeation tests with pure gases (He, N_2, CO_2 and CH_4) have been performed to characterize the membranes in terms of the N_2 permeating flux and ideal selectivity at different temperature and pressure conditions, varying from 400 to 600 Â°C and from 2.0 to 60 bar, respectively. Moreover, scanning electron microscopy (SEM) and an electron microprobe analyzer (EMPA) have been used to investigate the effect of the different operating conditions on the membrane surface. From the experimental results, all of the membranes show infinite selectivity towards N_2 permeation at each operating condition investigated and V showed better performance with respect to Nb, Ta, and Fe.

The V membrane is housed in the membrane reactor where a H_2 stream is used as the sweep gas to promote the NH_3 reaction. Lower pressure than the conventional Haber-Bosch process is used. Specifically, an operating pressure of 60 bar, instead of 200 bar (Haber-Bosch process), and a temperature of 500 °C are used as operating conditions to produce NH_3. The performance of the membrane reactor in terms of NH_3 conversion is, hence, investigated.

Copyright © American Institute of Chemical Engineers. All rights reserved.

485442 Capacitive Deionization of Brackish Water for Irrigation and Energy Storage

Abstract:

Irrigated agriculture water represents the bulk of demand for water in most of the countries suffering from acute water scarcity in the Near East and North Africa, as well as Mexico, Pakistan, South Africa, and large parts of China and India. Greenhouses are emerging solution. Current testing in Mozambique and Sierra Leone has demonstrated significantly reduced water use. Yet even a few gallons per day per greenhouse is challenging because many local water sources are brackish. Capacitive deionization (CDI) is a sustainable, energy efficient, and cost effective technology for removal of low or moderate salinity typical of brackish water as compared to technologies that remove the water instead. Indigenous production of the carbon for the electrodes holds the key to such technology being adopted as a source of irrigation water, simultaneously facilitating the use of such using renewable energy sources. Biochars have recently been explored as electrode materials in CD, and can be easily produced from local plant-based agricultural waste materials.

The goal of this study is thus to identify biomass process conditions to enable the best performance as electrode material in CDI. Studies on CDI so far have been targeted towards achieving the best adsorption capacity, which directs towards the use of highly processed and tailored carbons. This is mostly achieved through chemical activation processes using toxic or corrosive and expensive materials which require deionized water for subsequent washing of the carbon, for achieving better CDI performance. Hence, with the effort directed towards making the process viable for use locally in regions with acute water scarcity, a change in the approach to tailor the electrode material for use in CDI is proposed. This approach is to determine the best possible material utilization by ensuring maximum possible contact between the liquid and the biochar surface (wettability) and by generating the least Ohmic loss thus resulting in highest electrical efficiency.

Copyright © American Institute of Chemical Engineers. All rights reserved.

503648 The Last Quarter Century of Methods and Software for Algorithmic Process Synthesis

Abstract:

Process-Network Synthesis or PNS in short formally defined by Friedler and Fan in 1991 is the constructive procedure of building up complex processes or networks from elementary building blocks in order to achieve overall targets by the help of external and internal resources. Advanced algorithmic methodologies and computational tools developed in the last quarter of century could overcome the combinatorial explosion caused by the large number of alternative combinations of process elements having the potential to achieve the desired overall target facing industrial size problems. Moreover, aiming at finding the global optimum multiple component problems of the process-synthesis traditionally examined sequentially become solvable in integrated way, including reaction path synthesis, separation-network synthesis, heat integration, control design, and scheduling. The presentation aims at reviewing the related methods and software. Latest developments and future challenges are to be highlighted as well.

Copyright © American Institute of Chemical Engineers. All rights reserved.

492362 Near-Term and Sustainable Carbon Dioxide Removal: Is Bio-Energy with Carbon Capture and Storage (BECCS) the Right Answer?

Abstract:

Bioenergy with CO_2 Capture and Storage (BECCS) is commonly regarded as a vital yet controversial negative emissions technology (NET) to meeting our climate goals (Fuss et al., 2014). In this contribution we present a whole-systems analysis of the BECCS value chain associated with the cultivation, harvesting, transport, pre-treatment and conversion in dedicated biomass power stations in conjunction with CCS, of a range of biomass resources â€" both dedicated energy crops (miscanthus, switchgrass, short rotation coppice willow), and agricultural residues (wheat straw). We explicitly consider the implications of sourcing the biomass from different regions, climates and land types. The water, carbon and energy footprints of each value chain were calculated, and their impact on the overall system water, carbon and power efficiencies were evaluated. An extensive literature review was performed and a statistical analysis of the available data is presented. In order to describe the dynamic greenhouse gas balance of such as system, a yearly accounting of the emissions was performed over the lifetime of a BECCS facility, and the carbon "breakeven time" and lifetime net CO_2 removal from the atmosphere were determined. The effects of direct and indirect land use change were included, and were found to be a key determinant of the viability of a BECCS project. Overall we conclude that, depending on the conditions of its deployment, BECCS could lead to both carbon positive and negative results. The total quantity of CO_2 removed from the atmosphere over the project lifetime and the carbon breakeven time were observed to be highly case specific. This has profound implications for the policy frameworks required to incentivise and regulate the widespread deployment of BECCS technology. The results of a sensitivity analysis on the model combined with the investigation of alternate supply chain scenarios elucidated key levers to improve the sustainability of BECCS: 1) measuring and limiting the impacts of direct and indirect land use change, 2) using carbon neutral power and organic fertilizer, 3) minimising biomass transport, and prioritising sea over road transport, 4) maximising the use of carbon negative fuels, and, 5) exploiting alternative biomass processing options, e.g., natural drying or torrefaction. A key conclusion is that, regardless of the biomass and region studied, the sustainability of BECCS relies heavily on intelligent management of the supply chain (Fajardy & Mac Dowell, 2017).

References

Fuss, S., Canadell, J. G., Peters, G. P., Tavoni, M., Andrew, R. M., Ciais, P., Jackson, R. B., Jones, C. D., Kraxner, F., Nakicenovic, N., Le Quere, C., Raupach, M. R., Sharifi, A., Smith, P. & Yamagata, Y. (2014). Betting on negative emissions. Nature Climate Change, 4 (10), 850â€"853. DOI:Â 10.1038/nclimate2392

Fajardy, M., & Mac Dowell, N. (2017). Can BECCS deliver sustainable and resource efficient negative emissions? Energy and Environmental Science. DOI: 10.1039/C7EE00465F

Copyright © American Institute of Chemical Engineers. All rights reserved.

499278 A Physical Input-Output Model for the Food-Energy-Water (FEW) Nexus in Indiana

Abstract:

One of the major obstacles to understand the challenges of managing the production of food and energy while maintaining water availability and resources is the lack of physical insights into the regional economies. Physical Input-Output Table (PIOT) based models provide insights into physical interaction of various production sectors while taking advantage of structural information about the economy, thus quantifying the interdependency of food and energy production [1]. Furthermore, the environmentally extended form of PIOT allows the studying of environmental impacts from production systems by tracking the emissions to water and air by each production system. While few fully developed physical input output tables have been produced, their application has been demonstrated in measuring land footprints, analyzing the structure of the economy, and testing the resiliency of supply chains, all of which are important in FEW systems [2,3,4]. Hence, we developed an initial PIOT model for agro-based industries in Indiana to analyze the interdependency of food and bioenergy production in the state. In particular, the unidimensional flow of nitrogen was tracked as measure of production from various sectors, since N forms the basis of food production and the optimization of its use is critically important to food systems and ecosystem management. Indiana's agrobased economy is based on corn and soybean production, which together make up over 76.5% of Indiana's agricultural land [5]. Bioenergy is produced as both determining and co-product from the corn processing sectors. To build this model, we generated process models for each production unit and used the results to generate the PIOT as shown in [6]. For agricultural systems, we performed mass balances on production systems. An algorithmic approach was followed to find the statistical sources for information where available, as well as approximation methods from peer reviewed literature. Once the mass balances were completed they served as the basis for process models that fit into the PIOT building approach described in [6]. The structure of PIOT in Indiana is compared with an earlier PIOT developed for Illinois [7].

[1] Hoekstra, R., & van den Bergh, J. C. (2006). Constructing physical input–output tables for environmental modeling and accounting: Framework and illustrations. Ecological Economics, 59(3), 375-393.
[2] Hubacek, K., & Giljum, S. (2003). Applying physical input–output analysis to estimate land appropriation (ecological footprints) of international trade activities. Ecological economics, 44(1), 137-151.
[3] Altimiras-Martin, A. (2014). Analysing the structure of the economy using physical input–output tables. Economic Systems Research, 26(4), 463-485.
[4] Nakatani, J., Tahara, K., Nakajima, K., Daigo, I., Kurishima, H., Kudoh, Y., Matsubae, K., Fukushima, Y., Ihara, T., Kikuchi, Y. and Nishijima, A. (2017). A graph theory-based methodology for vulnerability assessment of supply chains using the life cycle inventory database. Omega.
[5] United States Department of Agriculture National Agricultural Statistical Service. (2017). 2016 State Agriculture Overview: Indiana. Retrieved from https://www.nass.usda.gov/Quick_Stats/Ag_Overview/stateOverview.php?state=INDIANA.
[6] Wachs, L. and Singh, S. (2016, November). Computational Approaches in Systems Modeling for Environmental Impacts of Industries Automating Physical Input-Output Tables (PIOTs) Via Aspen Process Modeling. Poster session presented at the 2016 AIChE Annual Meeting, San Francisco, CA.
[7] Singh S, Compton J, Hawkins, T, Sobota D, Cooter E., "A Physical Input-Output Model for Illinois" In Review

Copyright © American Institute of Chemical Engineers. All rights reserved.

501390 Interventions for Reducing Energy Impacts of Water Embodied in Domestic Food Trade: A Network Perspective

Abstract:

Integrated management of food, energy, and water resources has been advocated to mitigate unintended consequences of managing them in isolation. Food, energy, and water (FEW) systems are highly interdependent, interconnected, and exhibit numerous interactions at distinctive spatial scales. Sustainable management and availability of these resources will require quantitatively mapping their complex interdependencies and understanding their ecology. An integral component of food security and diversity involves trade of food items. Food trade also presents a pathway for transferring embodied resources and emissions associated with food production. For the United States (U.S.), irrigation adds significant value to food and feed production and represents a critical link for studying domestic FEW systems. Specifically, irrigation water withdrawals account for second largest freshwater withdrawals in the U.S. and has substantial associated energy and environmental impacts. We recently developed a network model of interstate food transfers for the U.S. The model was utilized to understand the energetic and greenhouse (GHG) impacts of irrigation embodied in interstate food transfers [1]. This analysis highlighted the high energy and GHG emissions embodied in irrigation associated with domestic food trade. We also proposed relevant network metrics to understand the structure of domestic food trade network and its implications for resilience.

In this work, we propose and evaluate a range of intervention strategies to enhance environmental sustainability and resilience of the domestic food trade network as they relate to irrigation and its embodied energy impacts. We use publicly available datasets to develop weighted and directed networks of interstate food trade and associated environmental impacts. We consider interstate food transfers of over 50 commodities including cereal grains, milled grains, livestock, and meat products. From a network perspective, the weighted food trade network consists of a few key states (Texas, California, Kansas) that control high throughput. We also observe that these influential states depend on water resources facing depletion risks (Colorado river basin, Ogallala aquifer), employ energy and water inefficient irrigation systems, and utilize electricity and diesel-based pumps. We evaluate the impact of technological, behavioral, and policy interventions using our interstate food network model to study their impact on the sustainability and resilience of the domestic FEW nexus. The investigated interventions include changes in irrigation practices, energy mixes, adjustments in diet patterns, and spatial changes in food production patterns. These strategies help in prioritizing and shedding light on interventions that aid in improving sustainability and resilience of the FEW nexus from a systems perspective as opposed to those that offer incremental advantages. The implications of these findings including their potential and limitations for enhancing the sustainability and resilience of the FEW nexus will be described in detail.

Reference Cited:

N. Vora, A. Shah, M.M. Bilec, and V. Khanna. Food-Energy-Water nexus: quantifying embodied energy and GHG emissions from irrigation through virtual water transfers in food trade, *ACS Sustainable Chemistry and Engineering,*

2017, 5(3), 2119-2128.

Copyright © American Institute of Chemical Engineers. All rights reserved.

500119 A General Framework for Process and Utility Networks Synthesis

Abstract:

There are sub-systems in chemical plants that consist of a number of sinks with demand for a certain property and a number of alternative sources that can satisfy these properties to fulfil the sink demands. Examples of such sub-systems are heat exchanger networks [1], fuel gas networks [2], water networks [3], utility networks [4], etc. Properties that are exchanged can be energy in the form of heat as in heat exchanger networks, or in both work and heat form as in the utility networks. Although each subsystem has attracted significant attention, each synthesis problem requires a new superstructure to be postulated whenever a new problem is encountered. Similarly, separation network systems, such as membrane networks [5] and distillation sequences [6], have been a focus of attention, yet a new superstructure representation is needed for each problem. A unified superstructure representation for these different classes of synthesis problems will be highly advantageous. Recently, a systematic framework for process intensification based on building-block superstructure has been proposed based on the idea of dissecting various unit operations into fundamental building blocks [7]. The proposed method can be also used for automatic flowsheet generation.

In this work, we extend the approach to provide a general framework for the synthesis of different networks using a similar representation using building blocks. Here, each block has feed and product streams ensuring interaction with the surroundings and can be viewed as sources and sinks, respectively. Also, streams between blocks are equipped with compressors or expanders depending on the source and sink block pressures. Mass and energy flow within the superstructure are facilitated via block material and energy balances and heating and cooling requirements are decided accordingly. Separation operations, such as gas membrane separations, are represented via two neighboring blocks separated by a semi-restricted boundary. The decisions over the assignment of boundary types between blocks are enabled via use of decision variables. By using all of these features of the building block superstructure, several different process network synthesis problems can be accommodated in a single superstructure with n and m number of blocks in row and column respectively. It should be noted that for different network synthesis problem, the block superstructure size could be reduced to improve the computation performance, i.e, using $2\times m$ superstructure to represent a pooling problem without intermediate pools. The overall process network synthesis problem is formulated as a mixed-integer nonlinear optimization (MINLP) problem where the nonlinear terms are due to splitting, energy balances and/or work calculations and discrete variables are for assigning alternative operations to each block and boundary in the superstructure. In this work, we will demonstrate that the proposed method is capable of unifying the superstructure representations for different class of process network synthesis problems via a range of literature problems.

References:

[1] Yee, T.F., Grossmann, I.E. and Kravanja, Z., 1990. Simultaneous optimization models for heat integrationâ€"I. Area and energy targeting and modeling of multi-stream exchangers.Â *Computers & chemical engineering*,Â *14*(10), pp.1151-1164.

[2] Hasan, M.F., Karimi, I.A. and Avison, C.M., 2011. Preliminary synthesis of fuel gas networks to conserve energy and preserve the environment. Industrial & Engineering Chemistry Research, 50(12), pp.7414-7427.

[3] AhmetoviÄ‡, E. and Grossmann, I.E., 2011. Global superstructure optimization for the design of integrated process water networks. AIChE journal, 57(2), pp.434-457.

[4] Elia, J.A., Baliban, R.C. and Floudas, C.A., 2010. Toward novel hybrid biomass, coal, and natural gas processes for satisfying current transportation fuel demands, 2: Simultaneous heat and power integration. Industrial & Engineering Chemistry Research, 49(16), pp.7371-7388.

[5] Alnouri, S.Y. and Linke, P., 2012. A systematic approach to optimal membrane network synthesis for seawater desalination. *Journal of membrane science*, *417*, pp.96-112.

[6] Giridhar, A. and Agrawal, R., 2010. Synthesis of distillation configurations: I. Characteristics of a good search space. Computers & chemical engineering, 34(1), pp.73-83.

[7] Demirel, S.E., Li, J., Hasan, M.M.F. Systematic Process Intensification using Building Blocks. Computers & Chemical Engineering, http://dx.doi.org/10.1016/j.compchemeng.2017.01.044.

Copyright © American Institute of Chemical Engineers. All rights reserved.

AUTHOR INDEX